Lecture Notes in Engineering

The Springer-Verlag Lecture Notes provide rapid (approximately six months), refereed publication of topical items, longer than ordinary journal articles but shorter and less formal than most monographs and textbooks. They are published in an attractive yet economical format; authors or editors provide manuscripts typed to specifications, ready for photo-reproduction.

Lecture Notes in Engineering

Edited by C. A. Brebbia and S. A. Orszag

69

T. K. S. Murthy
C. A. Brebbia (Eds.)

Advances in Computer Technology and Applications in Japan

Springer-Verlag
Berlin Heidelberg New York
London Paris Tokyo
Hong Kong Barcelona Budapest

Editors
Thiruvalam K. S. Murthy
Carlos A. Brebbia

Wessex Institute of Technology
Ashurst Lodge, Ashurst
Southampton SO4 2AA
United Kingdom

ISBN-13:978-3-540-54072-4 e-ISBN-13:978-3-642-84514-7
DOI: 10.1007/978-3-642-84514-7

Typesetting: Camera ready by author

61/3020-543210 Printed on acid-free paper.

PREFACE

This book contains chapters written by eminent scientists on the latest development in computer technology and applications in Japan.

The material contained in this book is an edited version of the presentation at a seminar arranged by the Computational Mechanics Institute of the Wessex Institute of Technology at Southampton during 1989. The objective of the Seminar was to provide an awareness of the considerable advances being made by Japanese scientists in the general area of information technology and in the so-called Fifth Generation Computer Systems.

In the first chapter, Watanabe of the NEC Corporation, describes advanced architecture and technology of supercomputing systems. This theme is followed by Nakamura of Tohoku University in the next chapter. Another type of supercomputer for vector processing, the FACOM VP 2000 Series is then described by Uchida of Fujitsu Ltd in Chapter Three.

Expert systems are presented in the next two chapters by Ueno and Oomari of Tokyo Denki University and by Koseki and Goto of the NEC Corporation. Important applications in computer graphics are described in Chapter Six by Ishii and Murakami of Fujitsu Laboratories. Hayashi from the same Laboratory then discusses neurocomputers in Japan. The final chapter by Noguchi of Tohoku University illustrates an important application in communications.

The Editors acknowledge the help of Computational Mechanics Publications in the preparation and sub-editing of the manuscript and are grateful for the high standards of presentation.

A proper balance has been maintained throughout the book by obtaining contributions both from industry and academia.

The Editors

CONTENTS

Advanced Architecture and Technology of Supercomputing Systems

T. Watanabe

Computer Engineering Division, NEC Corporation
1 - 10, Nisshin - cho, Fuchu, Tokyo 183, Japan

ABSTRACT

This paper presents a brief review of our vector and parallel processing experiences, basic hardware technologies for supercomputers and the architecture of the SX-3 Supercomputer with a vector peak speed of 22 GFlops for achieving the ultra high-speed processing. In particular, LSI and packaging technology, and the system and processor architecture of the SX-3 are described.

INTRODUCTION

The history of NEC's computer began in late 1950's. The first computer, called NEAC 1101, was designed for scientific and engineering calculations, and had floating-point arithmetic instructions with a speed of 300 Flops!

Since then, NEC entered into the computer business, and now is a total system supplier offering PC to supercomputers and all kinds of peripherals and terminals.

In 1983, we announced the Supercomputer SX System with a peak speed of 1.3 GFlops, and entered into the supercomputer business to meet the growing demands for large and high-speed computations. Just this April 1989, the follow-on SX System, the SX-3 series, was announced, which has a peak speed of 22 GFlops and is the first Japanese supercomputer employed a multiprocessor configuration.

This paper presents a brief review of our vector and parallel processing experiences, basic hardware technologies for supercomputers and the architecture of the SX-3 supercomputer for achieving the ultra high-speed processing.

REVEIW OF SUPERCOMPUTERS IN NEC

Our vector processing experience began with the NEC ACOS System 1000/IAP (Integrated Array Processor) announced in 1980. The ACOS System 1000 is a general purpose main-frame computer and was one of the fastest processors at that time. It featured an integrated vector processing functions called an Integrated Array Processor (IAP) which supported 60 vector instructions in addition to the conventional instruction set of the ACOS main-frame computer. Its vector peak speed was 14MFlops.

The basic vector architecture of the System 1000/IAP is memory-to-memory architecture shown in Figure 1, considering the familiarity with the conventional instruction set. One of the major characteristics of those instructions is that three different types of vector data with different strides can be specified in one instruction using data descriptors, as shown in Figure 1.

By fully utilizing these vector instructions, the Fortran compiler supported the automatic vectorizing functions although the vectorizable DO loops were limited to the simple DO loops at first. Some kinds of DO loops such as 'Search for Maximum/ Minimum' were vectorized by vector function calls.

Based on the expertise of the System 1000/IAP, the NEC Supercomputer SX series shown in Figure 2, which was the first supercomputer to break the 1 GFlops barrier, was developed to meet the growing demands for large and high-speed computations. Major features of the SX Series are:

- the fastest single processor in both vector and scalar processing,
- 16 multiple-parallel vector pipelines with gather/scatter and masked vector operations,
- RISC architecture with 128 scalar registers for high-speed scalar processing,

which were succeeded to the newest SX-3 series, and will be described later.

In addition, the new Fortran compiler with sophisticated vectorizing functions for the SX series was developed, where DO loops with the complicated statements such as the conditional IF statements, the intrinsic functions and search for max/min can be vectorized.

As for the parallel processing research and development in NEC, major activities are conducted in our research laboratory.
Examples of recent achievements are Cenju and HAL. Cenju is a highly parallel processing system which consists of 64 MC68020 microprocessors with floating-point accelerators and is designed for circuit analysis for semiconductor development. HAL, shown in Figure 3, is also a highly parallel processing system and an ultra high-speed simulation engine for the internal logic functions of computer system. And it was used for the development of supercomputer itself! In HAL, specially designed 32 processors are connected through multi-stage network consisting of router-cell LSI's.

HARDWARE TECHNOLOGIES FOR SUPERCOMPUTERS

One of the key factors for achieving high-speed processing particularly in a pipelined vector processor is to shorten machine cycle time as much as possible. For achieving high-speed machine cycle time, we need high-speed switching device and high-density packaging to reduce a propagation delay on wiring.

NEC has traditionally employed the CML (Current Mode Logic) LSI, similar to the ECL LSI, for large-scale main-frame computers and supercomputers as the high-speed switching device. Figure 4 shows a trend of gate switching speed of various device technologies. We can say that the switching speed of bipolar LSI's is still going faster.

In the supercomputer SX-3 series, we developed and employed the proven but highly advanced silicon CML technology as the basic LSI chips. The LSI chip, shown in Figure 5, has 20,000 gates with a gate switching speed of 70 picoseconds comparable to that of GaAs chip, but it is extremely higher in chip density than GaAs.

To mount those LSI's on the package, we have also traditionally employed the so-called Multi-Chip Package (MCP) technology as the basic packaging technology where the TAB (Tape Automated Bonding) LSI's are directly mounted on the ceramic substrate as shown in Figure 6. Figure 6 (a) shows the MCP for our first generation supercomputer SX series which has up to 36 LSI's on the ceramic substrate of the size of 10 cm x 10 cm. In the SX-3 series, the size of the ceramic substrate shown in Figure 6 (b) is enhanced to 22.5 cm x 22.5 cm on which 100 LSI's with 20,000 gates each can be mounted, resulting in a maximum of 2 million gates per package, and thus in higher gate density and shorter wiring length.

For the wirings among LSI's, two signal layers were sufficient for the former SX series MCP with 36 LSI's. In the SX-3 series MCP, the signal layers are augmented, but still only four layers are sufficient to connect all the wirings among 100 LSI's by ultra fine wiring patterns of 25 micron width. We employed the poly-imide organic material for insulation of inter-layers. Its major characteristics is low dielectric constant which leads to lower propagation delay of signals on the wiring pattern compared to inorganic material.
Table 1 shows a summary of hardware technologies employed in the SX-3 and SX series

Table 1 Hardware Technologies

		SX-3	SX-2A
Logic LSI	Speed Gates	70psec 20,000Gates	250psec 1,000Gates
Memory LSI	Speed Bits+Gates	1.6nsec 40KBits +7,000Gates	3.5nsec 1KBits
Main Memory RAM	Speed Bits	20nsec 256KSRAM	40nsec 256KSRAM
LSI Package	Size Chips	22.5cm x 22.5cm 100	10cm x 10cm 36

The denser the gate density in package, the more the heat density increases. In order to cool the package efficiently, the water-cooled system is employed in this packaging system as shown in Figure 7. The so-called cold plate, through which the cooled water is circulated, is in contact with the MCP where the heat generated in each LSI chip is conducted to the cold plate and cooled by water.

SYSTEM ARCHITECTURE OF THE SUPERCOMPUTER SX-3

Our basic approach to realize a high-speed computer system is to enhance a single processor performance to the ultimate, and then to combine those ultra high-speed processors constituting a multiprocessor system.

Figure 8 and 9 show a system appearance and a maximum configuration of the SX-3 system respectively, and Table 2 shows the system specifications together with those of the SX-2A, the top end model of the former SX series. In a maximum configuration, four arithmetic processors share a common main memory unit which has a capacity of up to 2 GBytes. The shared memory system and a small number of high-speed processors to realize a multiprocessor system give users the ease of use and easy programming environments, because they don't need to care about the memory allocation algorithm, different from the distributed memory system, and don't need to augment the degree of parallelism to fully utilize the hardware capability.

In Figure 9, the Control Processor (CP) is an extension of our main-frame computer ACOS system and performs supervisory functions such as job scheduling, resource management and I/O management. On the other hand, the Arithmetic Processor (AP), which has internally a scalar and vector unit, is an ultra high-speed Fortran engine and executes all the user codes compiled by Fortran compiler.
The Control Processor Memory (CPM) with a capacity of up to 256 MBytes is dedicated to the execution of the operating system functions and works also as a large I/O buffer. The Main Memory Unit (MMU) is a large and fast memory for the execution of programs running on the Arithmetic Processors. To transfer a large amount of vector data quickly and smoothly, the MMU is divided into a maximum of 1,024 independent banks, that is, 1,024 way interleaved system is employed, and is made up of 256 KBits, 20 nsec static RAM chips.

The Extended Memory Unit (XMU) is a large capacity semiconductor memory unit ranging from 1 GBytes to 16 GBytes, and works as a very high-speed virtual disk unit. The XMU, which has a transfer speed of 2.75 GBytes/sec, is used for temporary/ permanent disk files, disk cache buffer and job swapping files.

The SX-3 can configure up to four I/O processors with an aggregate transfer speed of 1 GBytes/sec. Each I/O processor has up to 64 channels through which various peripherals such as disk units, cartridge tapes, laser printers, optical disk units are connected.

Table 2 System Specifications of the SX-3 and SX-2A

MODEL	SX-3							SX-2A
	44	42	24	22	14	12	11	
Vector Peak Performance	22GFLOPS	11GFLOPS		5.5GFLOPS		2.75 GFLOPS	1.37 GFLOPS	1.3 GFLOPS
Number of Arithmetic Processors	4		2		1			1
Reg. Vector Registers	144KBx4	72KBx4	144KBx2	72KBx2	144KB	72KB	36KB	80KB
Reg. Vector Mask Registers	32		16		8			8
Reg. Scalar Registers	64Bitsx128x4		64Bitsx128x2		64Bitsx128			64Bits x128
Vector Pipelines/Sets	64/16	32/8		16/4		8/2	4/1	16/4
Cache Memory	64KBytesx4		64KBytesx2		64KBytes			64KBytes
Main Memory [Max]	2048MBytes			1024MBytes			512 MBytes	1024 MBytes
Extended Memory [Max]	16GBytes							8GBytes
I/O Channels [Max]	256							64
I/O Transfer Rate	1GBytes/Second							192MBytes /Second

PROCESSOR ARCHITECTURE OF TEH SX-3

Figure 10 shows the internal structure of the Arithmetic Processor (AP). The Arithmetic Processor is functionally divided into the Scalar and Vector Unit. In the maximum configuration, there are four identical sets of vector pipelines in the Vector Unit. Each pipeline set handles every fourth vector operation as shown in Figure 11, which can be said "Hardware Micro-tasking". The hardware of the Vector Unit controls the assignment of the vector operations to the corresponding pipeline set. This configuration, therefore, looks like one pipeline set speeded up four times compared to one set of the Vector Unit from the software view point. And each pipeline set consists of two pipeline adders and two multipliers, each of which can work independently in 2.9 nsec cycle time. Thus, it results in a vector peak speed of 5.5 GFlops in a single processor.

The Vector Unit has 144 KBytes of vector register to store intemediate results of vector operations, reducing memory access frequency and supplying vector data quickly to the vector pipelines. In addition, the Vector Unit can handle vector data with constant stride as well as the contiguously located vector. List or indirect vector located randomly can be also handled by scatter and gather operations.

DO loops with conditional IF statements can be vectorized by masked vector operations and vector mask registers. In addition, nested IF statements can be also vectorized by using 8 vector mask registers and logical operations on those vector mask registers. To handle vectorized conditional IF statements efficiently, Compress and Expand operations are supported.

The architecture of the Scalar Unit is based on the so-called RISC architecture where the high-speed scalar processing is achieved by simplified architecture and hardware control. Followings are major scalar features of the SX-3.

- 128 64-bit general purpose scalar registers for address calculations, logical, integer and floating-point operations,
- simplified three-operand instruction format to make flexible use of scalar registers,
- load/store architecture to simplify the hardware control,
- 64 KBytes of cache memory and 4 KBytes of instruction buffer with branch history,
- scalar arithmetic pipelines to increase concurrencies of executions.

Figure 12 shows major instruction formats, RX and RR/RV. RX format is 64 bits long, which is used for memory access instructions such as load, store and branch instructions. No arithmetic operation is supported in the RX format. RR/RV format is used for arithmetic instructions, where x, y and z fields specify registers or immediate data like Sx <— Sy op Sz, in which Sx, Sy and Sz are scalar registers specified in each corresponding field.

The Scalar Unit has 128 scalar registers which are used for any purposes and can be equally accessed by any instructions using scalar registers. A large amount of general purpose registers give the Fortran compiler greater flexibility for optimization and reduce memory access frequency and pipeline interlock due to the register conflicts.

For multiprocessor control, 3 KBytes of communication register are provided to make fast access and reduce memory contention. The so-called 'Test and Set' instruction is supported on these registers to synchronize the multiple tasks and control the exclusive use of shared resources.

CONCLUSION

In this paper, first of all, a brief history of vector processing and supercomputer was reviewed, and then the architecture and basic hardware technology of the newest supercomputer SX-3 for achieving ultra high-speed processing was presented.

It should be noted that our basic design approach for high-speed processing is to pursue the ultimate performance of a single processor, and then to constitute the multiprocessor system by combining a small number of those high-speed processors. We believe that the number of processors in multiprocessor system should be as small as possible from the point of view of ease-of-use. It should be also noted that one of major technological progresses, which contributed to the realization of this type of multiprocessor system, is in LSI and high-density packaging technologies.

REFERENCES

1. Furukatsu, T., Watanabe, T. and Kondo, R. Supercomputer SX System with a Peak Performance of 1.3 GFlops and 6 Nanosecond Cycle Time, Nikkei Electronics, 356, pp.237-272, Nov. 9, 1984. (Japanese)
2. Watanabe, T. Design Concept for Highspeed Vector and Scalar Processing: Architecture of the NEC Supercomputer SX System, in ICCD' 86, pp.38-41, Proc. of ICCD, 1986.
3. Watanabe, T. Architecture and Performance of NEC Supercomputer, Parallel Computing, 5, pp.247-255, 1987.

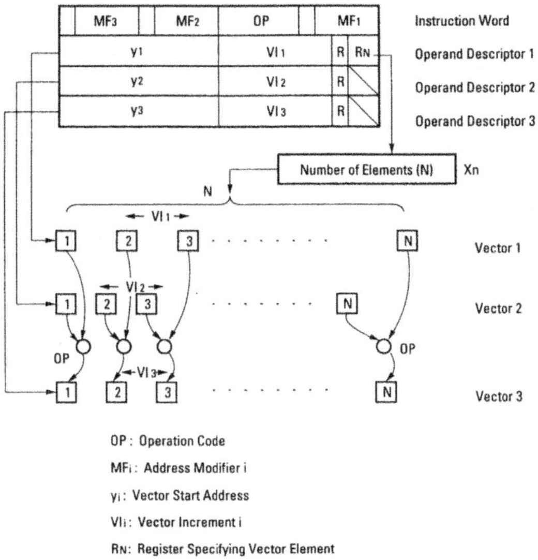

Figure 1 Vector Instructions of System 1000/IAP

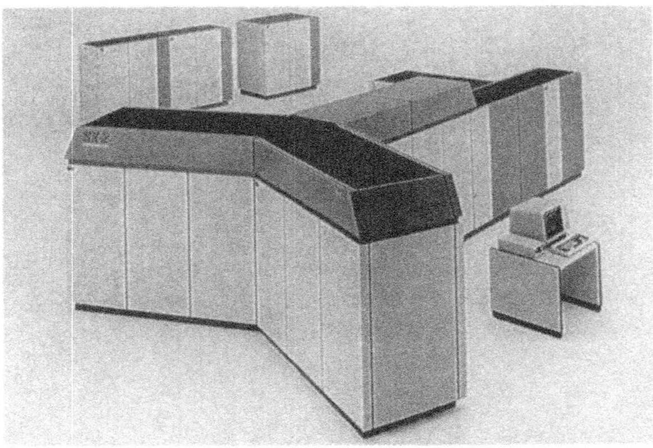

Figure 2 Supercomputer SX-2

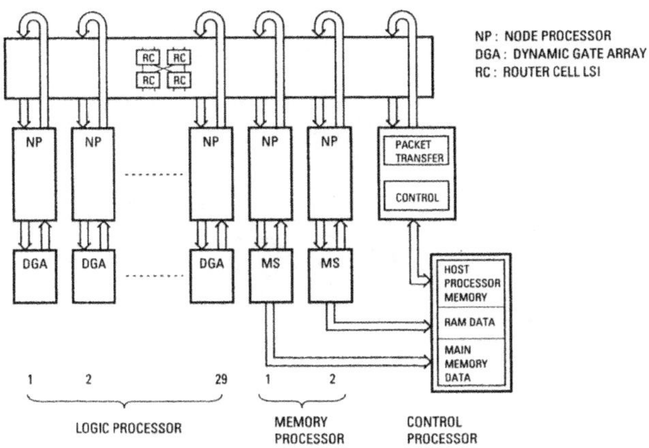

Figure 3 Configuration of HAL

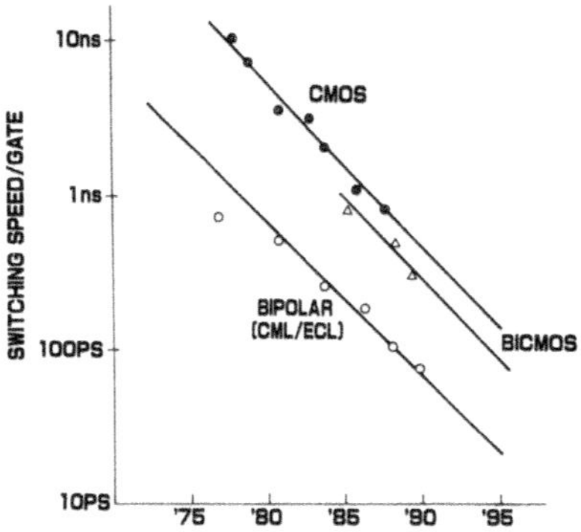

Figure 4 Trend of Various Device Technologies

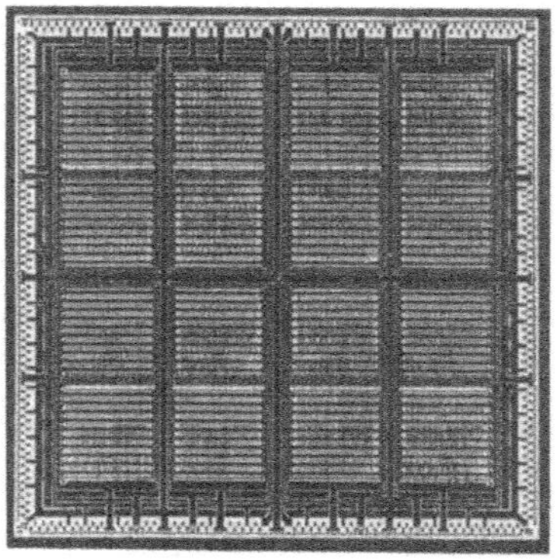

Figure 5 20,000 Gate CML LSI

(a) MCP for the SX-2

(b) MCP for the SX-3

Figure 6 Multi-chip Package (MCP)

Figure 7 Liquid Cooling Module for the MCP

Figure 8 Supercomputer SX-3

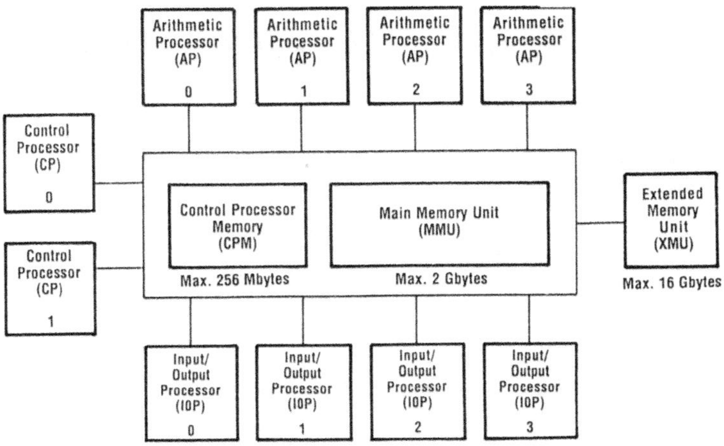

Figure 9 System Configuration of the SX-3

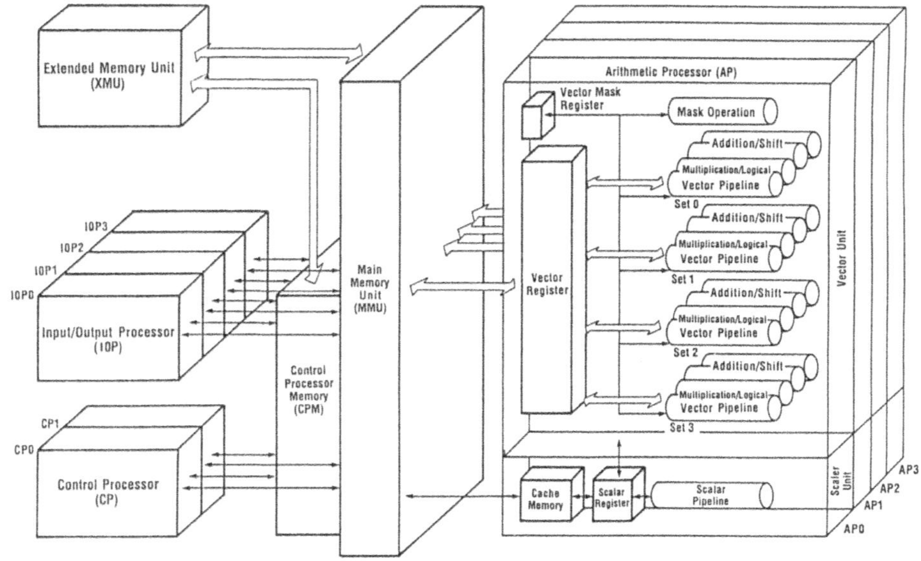

Figure 10 Configuration of the Arithmetic Processor

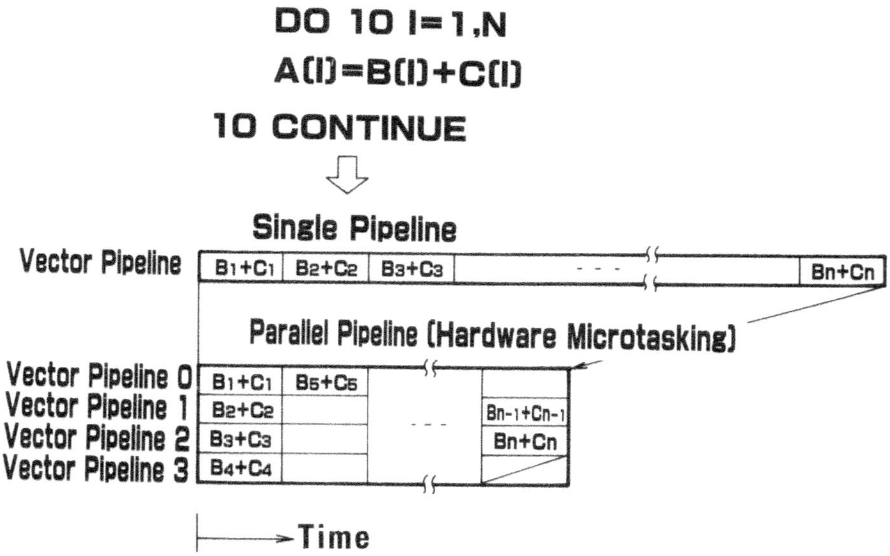

Figure 11 Execution of Parallel Pipeline

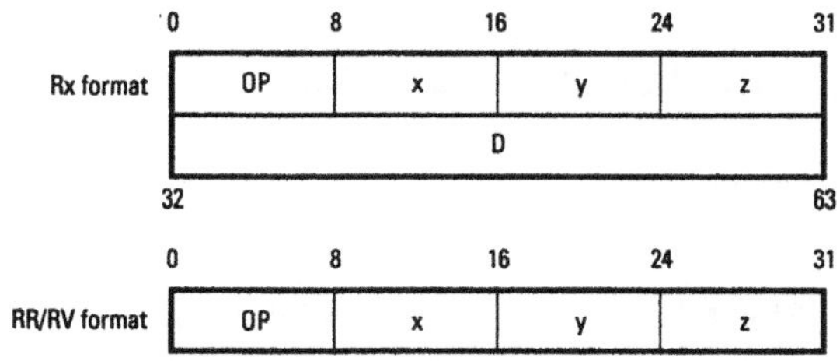

OP : Operation code

x,y,z, : Specifying the scalar register, vector register, or immediate data.

D : Specifying the displacement of the operand address.

Figure 12 Major Instruction Formats

Parallel and Supercomputer Architecture in Japan

T. Nakamura

Dept. of Mechanical Engineering, Tohoku University, Sendai 980, Japan

ABSTRACT

Parallel processing is the technology for speedup of computers. There are several ways of realizing it. According to the well-known Flynn classification of von Neumann machines into SISD, SIMD, MISD and MIMD types, parallel processing is commercially due to the SIMD and MIMD types. In this seminar, we focus on supercomputers (primarily in Japan) based on pipelining. Currently, Japanese supercomputers have a tendency to challenge to realize the peak performance using only multiple-parallel pipelining. We describe Japanese supercomputers with their advantages and disadvantages compared with the Cray series. In view of supercomputer architecture, though Japanese systems have features of speedup in linear data structures, there may be a few problems in their reliability, flexibility, and availability. In addition to improvement of throughput, these problems may be solved by means of multiprocessor system development. However, we are going to mention the other possibility of solving these problems in Japanese supercomputers by using an MISD concept.

I. INTRODUCTION

Supercomputing general is feasible in scientific and engineering field. Especially, numerical operations are in fashion at the moment. On the other hand, logical operations are necessary but not always sufficient in solving artificial intelligence problems (1).

This seminar on Parallel and Supercomputer Architecture in Japan will handle a concurrent technique of how to process in parallel and how to supercompute numerical/logical calculations. Further, we are going to present a numerical and symbolic vector supercomputers through the idea of general purpose pipelining.

First of all, parallel technique is mentioned using the conventional techniques. However, in view of supercomputers this will be summarized towards the supercomputing techniques. The advantage of using parallel processing is very strong to speedup the computation capability. Parallel techniques is composed of several techniques, such as multiprocessing, pipelining, and array technique. To realize supercomputers one or two of these techniques are used simultaneously in commercial supercomputer productions in Japan.

Implementation of supercomputing as vector machine is supported primarily by pipeline and array technologies or pipeline and multiprocessing ones (2). The former is used in Japanese supercomputers and the latter in the United States. Basically the Cray series in the US is the original of supercomputers and then this series is based on the computation scheme of pipelining and multiprocessing. This scheme is probably an ideal fashion in commercial style in view of computation capability, flexibility, availability, and reliability.

Section 2 mentions a fundamental scheme of parallel processing with the scheme of Cray series. The architectures of Japan's supercomputers will be presented in Section 3. Section 4 will serve the crisis of capability in solving every problems including numerical calculation and list processing. The paper will be summarized in Section 5.

II. PARALLEL PROCESSING ARCHITECTURE

In supercomputing, parallel processing is introduced as pipelining and multiprocessing technologies. These technologies are, in reality, usable and commercial in view of parallel technology general. Therefore, supercomputers have so far been supported by these technologies. Here we must review the pipelining and multiprocessing.

Multiprocessing has been realized by shared memory and message passing ways shown in Fig. 1 (3). In the case of shared memory, processors access other processors through memories via switch (network) as shown in Fig. 1 (a). In this case, multiprocessors are in tightly coupled MIMD architecture using shared memory among multiple processors, and called shared memory multiprocessors. On the other hand, in the case of message passing as in Fig. 1 (b), the system is called distributed memory multicomputers. This class corresponds to loosely coupled MIMD architecture with distributed local memories attached to multiple processor nodes.

Vector machines are constituted by pipelining or array of processing units as vector is generally handled in a SIMD architecture. Array of processing units has so far been considered to be suited partly but not always suited. Anyway depending on the problems and data, the array type is limited in using. The notion of array will be considered in thinking parallel pipelining of the same kind of arithmetic pipeline units.

The difference between vector machines and multiprocessors is how they can divide the data flow graph of the program (3). Firstly the graph is sliced to provide long vectors which are corresponding to the horizontal slicing in Fig. 2. Here, the data flow graph is partitioned into four tasks, each of which is executing on one processor. Within each tasks, there are some vectors of arithmetic that will be processed by vector machines. Thus, the graph is going to be processed by a vector machine and a multiprocessor.

When we think of supercomputers, we are forced to deal with shared memory multiprocessing supercomputers coupling suitable number of supercomputing units in terms of improving throughput, reliability, flexibility, and availability. Such supercomputers like Cray X-MP series are mainly produced in the US. Fig. 3 shows the system organization of the Cray X-MP-2 that has two processors (supercomputing units) (3). These are interconnected with each other using shared memory. Also, Fig. 4 illustrates the Cray X-MP-4 architecture including four supercomputing units (4). These Cray X-MP series are in architectural design based on vector processing and multiprocessing.

III. JAPAN'S SUPERCOMPUTERS

Japan's supercomputers are different from America's supercomputers. The former is based on pipelining and arrays using pipelines. The latter is due to pipelining and multiprocessing. Japan's supercomputers are arranged by Japan's government and its institute. Consequently, Japanese industrials tried to construct almost the same machine based on the same architecture. Its notion in designing supercomputers is from the chaining of pipeline operations. We have two modes in pipelining that mean independent vector operations that are performed separately from each other, and chained operations that are used to combine successive computations.

Japan has three large-scale supercomputers by Fujitsu, Hitachi, and NEC as in shown in Fig. 5. As we described previously, these companies' machines are produced under the same design philosophy. Especially, using parallel pipelines, which is regarded as arrays of pipelines, these machines can operate chaining of numerical operations. Vector registers are used to intercommunicate with pipelines, where these pipeline are used successively in the same kind of pipelines or different kind of pipelines. Vector registers are of course faster than main memory, so that it is profitable to realize chaining among operations within arithmetic pipelines.

According to the Alphabet order of each company's name, we firstly describe the details of Fujitsu's machine VP-200 in Fig. 6 (4). So far this company have produced the VP series. Basically Fujitsu has been developing software and applications of the series. Applications analysis is generally very important to realize real supercomputers. In this meaning, Fujitsu is very active to prepare many kinds of scientific applications and business applications.

Hitachi is rather thinking about speedup of its series in view of powerup of logic design. Fig. 7 displays the Hitachi S-810/20 architecture (4). Of course, Hitachi is developing their software for the supercomputers.

NEC has so far constructed very fast machine SX-2 which advanced a speedup of supercomputers. Fig. 8 portrays the SX-2 system configuration based on the pipelines' arrays called a multiple-parallel pipeline scheme with a vector register (5). Also, in Fig. 9 we mention the various pipelines of the SX-2 (5).

In order to improve the execution of a program, it is essential to raise the vectorizing ratio of programs. For example, in SX series there are hardware functions, for improving the vectorizing ratio, such as masked vector arithmetic function for vectorizing loops with IF statements, vector compression and expansion functions, basic arithmetic functions, etc (5). Fig. 10 shows these functions.

IV. SUPERCOMPUTING LIST-STRUCTURED DATA (6)

"What is architecture?" is the rigid problem in computing systems. A notion of this problem is established in a paradigm for application-driven parallel processing research (7). This is to select candidate applications, develop parallel models of computation for them, discover their parallel algorithms, and then solve the parallel language expressiveness and implementation efficiency problems separately (8).

Supercomputers have mainly been developed in the area of numerical analyses. As the data structure is mainly an array (including a vector), pipelining is adopted to realize high-speed data execution. This pipelining has been regarded as the parallel processing in view of parallel architecture. On the other hand, multiprocessor systems have the most flexible computer architecture in exploiting arbitrarily structured parallelism. Thus, pipelined multiprocessors represent state-of-the-art design in supercomputers, and many supercomputer manufacturers are adopting pipelining in upgrading their high-end models (2).

In this section, we aim at processing list-structured data of objects of function-level programs (9) on a supercomputer as an example of applications. The supercomputer used is composed of a general-purpose pipeline which has been proposed by one of the authors (10).

IV-I. GENERAL-PURPOSE SUPERCOMPUTER

Supercomputers are crucial to many scientific, technological, and engineering applications. Various applications including not only numerical analyses but also large databases and knowledge engineering will require general-purpose supercomputers (11). In this section, we mention the organization of a general-purpose supercomputer directed towards the objectives outlined above. Both pipelining and architecture of supercomputers may soon relate naturally with respect to the flowing data streams in pipelines.

It is clear that the generality of processing in pipelines seems to be unnatural in view of processing time in each segment. The definition of "general-purpose" is based on the various applications, for example, related to list-structured data (12). Of course, numeric calculations, like usual computations, will be included in the category of the "general-purpose," and will be executed in the same supercomputer. In this article, the key to a general-purpose supercomputer is presented at the architecture levels. Fig. 11 shows the scheme of how to execute general-purpose supercomputing in the pipeline. The key to processing is based on the function-changing segment in the pipelines. We should point out that data is primary and its program is secondary. The ways to change functions of each segment are considered in terms of the architectural philosophy even though the methods are limited in pipelining. According to the data processing scheme in Fig. 11, this pipeline can be regarded as an MISD architecture.

To implement the notion shown in Fig. 11, we must consider how to produce a general-purpose pipeline. As data is processed by its own program, it seems that the program tags along with its data. However, the transmission of the data and program require the bus of wide bandwidth. In terms of VLSI technology, such a bus transfer is not the optimum way. Under pipeline scheme, we arrange several processors in pipeline fashion as a multiprocessor system whose segments share memory including the structured data. Fig. 12 depicts the overview of the general-purpose pipeline based on multiprocessing, where data is stored in the structured-data memory and program flows through the pipeline composed of several processors while processed. Consequently, the structured data is processed in a pipeline fashion.

The processing elements (PEs) of the pipeline machine is proposed to be one in Fig. 13. The meaning of each structure element of this PE can be interpreted with respect to the usual notion of supercomputers. MAU is a memory access unit to the structured-data memory. FU means a function unit where arithmetic and logic operations are carried out. LU assists in storing the information for linkage between processes of tasks. IU is an interpreting unit which interprets the function-level programs going through the PE. IC is for control of receiving the program and pointers pointing out the structured data. OC is for control of transmitting the program and pointers.

IV-II. A WAY TOWARD SUPERCOMPUTING FOR VARIOUS APPLICATIONS

This section indicates how to execute list-structured data as an example of some of the various applications. The list-structured data is not always for various applications but rather has generality in recursively representing the connections of data. In order to perform pipeline execution of such list-structured data included in objects of FP programs, recursion of a function (program) can be introduced. The reason is described as follows. A recursive function is analyzed (rather unfolded) into instances at some or more levels to be piecewise processes, and then is synthesized (rather folded). A recursive function is well suited to test the scheme of pipelines since the processing time of each instance is almost the same.

The structure of the recursive functions is classified into tree-structured and linear-structured types in their expansion into some instances with levels each. In view of proper example of pipelining and for simplicity, let us deal with the linear-structured recursive function. From the limitation of pipelining we must try to simultaneously execute the unfolding and folding of the recursive processing. This means that while executing the unfolding, the folding should/can be done in each segment of a pipeline. Fortunately, in the case of FP programs, the programs apply to one object as one argument, and can be composed of many elements as a sequence. This corresponds to the linear-structured recursion.

The folding process is in parallel with the unfolding process by using the notion of "Pipeline cons" which is stemmed from the "Lenient cons" to realize the parallelism in dataflow computers (13). This pipeline cons is interpretively shown in Fig. 14 (a). To make parallelism in creating one cell of list-structured data and in writing one pointer into the car field of the cell and the other pointer the cdr field, cons is basically decomposed into the three parts: "getcell," "writecar," and "writecdr," respectively. In Fig. 14 (a), instead of the getcell, writecar, and writecdr, the abbreviated form "gcell," "wcar," and "wcdr" are used for simplicity. Here, the getcell means providing one new cell, the writecar means writing down a pointer into the car field of the cell, and the writecdr means writing down a pointer into the cdr field of the cell. In pipeline cons, when getcell and writecar are provided in one instance at a level i, the writecdr is executed in the instance at the one lower level i-1. And then a pointer at writecdr in the instance at the level i-1 is written down into the cdr field of the cell in the instance at the level i. Such a process is successively continued as shown in Fig. 14 (b).

Before executing function-level programs under the above pipeline cons, we had better review some of the rules of the function-level language (9),(14).

Function-level programs consist of objects, functions, functional forms, definitions, and one operator called "application."

Objects are numbers, symbols, words, or sequences. A sequence $<x_1, x_2, \ldots, x_n>$ of objects consists of x's which are either numbers, symbols, words, or sequences.

Primitive functions transform one object into another. As an example, there exist selector functions, arithmetic functions such as +, *, etc., transpose, and distribution functions.

Function definitions define new functions in terms of old ones.

Application is the operation of applying a function to an object.

Functional Forms are expressions denoting programs, which are built from existing programs using program-forming operations (PFOs). We mention some examples of these PFOs as follows:

a. Composition of f and g:
   $(f \circ g):X = f:(g:x)$
b. Construction of $f_1, f_2, \ldots, f_n$:
   $[f_1, f_2, \ldots, f_n]:x = <f_1:x, f_2:x, \ldots, f_n:x>$
c. Condition (of p, f and g):
   $(p \rightarrow f;g):x = f:x$ if $p:x$ is true,
   $= g:x$ if $p:x$ is false

d. Constant of an object y:
   y:x = y for any x, the function y gives the result y
e. Insert of f:
   $/f{:}\langle x_1, x_2,\ldots, x_n\rangle = f{:}\langle x_1,/f{:}\langle x_2,\ldots,x_n\rangle\rangle$
f. Apply-to-all of f:
   $\alpha f{:}\langle x_1, x_2,\ldots,x_n\rangle = \langle f{:}x_1, f{:}x_2,\ldots, f{:}x_n\rangle$

The FP program for the inner product is regarded as the proper one to explain the execution process of our pipelining and does not lose generality in list-structured data processing. The program is as follows:

Define Inner Product = (insert +)∘(apply-to-all *)∘ transpose

Or in abbreviated form:

Def IP = $(/+)\circ(\alpha *)\circ$trans

Focusing on the trans and considering two vectors <1,2,3>, <4,5,6> each of which is an element of a sequence as an object applied to the defined function IP, we show the process of recursive execution of the function IP. The function of the trans is to make a transposition:

trans: <<1,2,3>,<4,5,6>> = <<1,4>,<2,5>,<3,6>> (1)

In terms of recursive expressions, (1) is transformed into the following expression.

trans = null 1 → []; apndl [α1, trans αtl] (2)

"1" is one of selector functions, which chooses the first element of a sequence:

$1{:}\ \langle x_1,x_2,\ldots,x_n\rangle = x_1$

null 1 checks whether or not the first element of a sequence is null (NIL). [] is consequently the function to generate an empty sequence when the result of the above condition test is "true." α1 means picking up the first elements from sequences to be the elements of a given (old) sequence to create the first element to be a new sequence when the result of condition test is not NIL. Further αtl means retaining, in this case, two sequences whose elements each are the remainder of each sequence in a given sequence as an object.

When (2) is applied to <<1,2,3>,<4,5,6>>, the unfolding of this recursive function IP is executed according to [α1, trans αtl] in the second part of condition in this function. Then, the folding of this function is executed by executing "apendl" of the second part until NIL appears according to the first part in (2). This scheme is interpreted by using Fig. 15 (a) which shows list-structured data composed of the two vectors <1,2,3>, <4,5,6>. According to the pipeline cons and pipelining, (2) will firstly work as an unfolder which generates the new vector <1,4> from the two old vectors <1,2,3>, < 4,5,6>. And then the unfolder creates one cell by the getcell "gcell," and writes the pointer for the vector <1,4> by the writecar "wcar" in the first instance of (2). Next, the address of the cell created in the first instance is transmitted into the next instance, and then the work to write down the pointer for <2,5> in the cdr field of the cell is carried out by a folder. Fig. 15 (b) displays the new list-structured data for the three vectors <1,4>, <2,5>, <3,6>. This process is continued until finding out NIL, based on the pipeline cons to realize the parallelism between unfolding and folding.

In view of the general-purpose pipelining, the process of executing "trans" to the two vectors <1,2,3>, <4,5,6> is demonstrated in Fig 16. Especially, the unfolding and folding of the recursive function (2) is interpretively shown in terms of the implementation of its function.

Usually, the number of segments of this pipeline described above is not always sufficient enough to execute the whole program, because the number of segments of this pipeline is limited in view of its fabrication problems (cost etc.). In this case, under the notion of serialization, we have to consider how to partition a program into some tasks to be executed in a locally parallel computation fashion (15). Let us consider, for example, the case where we apply the function "transpose" to an object consisting of two relatively longer vectors of length n on the pipeline composed of m segments (n > m). The serialization is necessary when long vectors' computations are carried out based on locally parallel computations by m segments. This computing scheme will be implemented according to the architectural design like the loop-style pipeline (16).

V. CONCLUSION

In this seminar, we have focussed on Japan's supercomputers based on pipelining as we compare these supercomputers with the Cray series in view of parallel architecture. Though Japan's systems have features of speedup in linear data structures, there may be a few problems in their reliability, flexibility, and availability. We have mentioned the possibility to realize supercomputing of list-structured data under high speed operation by using an MISD concept.

REFERENCES

(1) L. Uhr, "Multi-computer architectures for artificial intelligence," Wiley Interscience, 1987.
(2) K. Hwang, "Multiprocessor supercomputers for scientific/engineering applications," Computer, vol. 18, no. 6, pp. 57-73, June 1985.
(3) V. M. Milutinovic', "Computer Architecture," North-Holland, 1982.
(4) K. Hwang and D. Degroot, "Parallel Processing for super-computers and artificial intelligence," Mc Graw-Hill,1989.
(5) T. Watanabe, T. Furukatsu, R. Kondo, T. Kawamura and Y. Izutani, "The supercomputer SX system : An overview," Proc ICS'87 supercomputing vol. 1, pp 51-56, 1987.
(6) T. Nakamura, Y. Takai and T. L. Kunii, "Pipelined supercomputing with list-structured data," Proc ICS'87 supercomputing vol. 1, pp 410-415, 1987.
(7) T. L. Kunii, "Application development systems," Springer-Verlag, 1986.
(8) P. C. Patton, "Multiprocessors: Architecture and applications," Computer, vol. 18, no. 6, pp. 29-40, June 1985.
(9) J. Backus, "Can programming be liberated from the von Neumann style? A functional style and its algebra of programs," Comm. ACM, vol. 21, pp. 613-641, Aug. 1978.
(10) T. Nakamura, "Software of the brain structured computer," Proc. of The IEEE Eighth International Computer Software & Applications Conference, pp. 408-414, Nov. 1984.
(11) L. A. Zadeh, "Coping with the imprecision of the real world," Comm. ACM, vol. 27, pp. 304-311, Apr. 1984.
(12) T. Nakamura, K. Sakai and Y. Mishina, "Function-level computing on the brain structured computer," Proc. of the IEEE Ninth International Computer Software & applications Conference, pp. 90-96, Oct. 1985.
(13) M. Amamiya, M Takesue, R. Hasegawa and H. Mikami, "Implementation and evaluation of a list-processing-oriented data flow machine," The 13th Annual International Symposium on Computer Architecture Conference Proc., pp. 10-19, June 1986.
(14) J. Backus, "Function-level computing," IEEE Spectrum, vol. 19, no. 8, pp. 22-27, Aug. 1982.
(15) A. Iizawa and T. L. Kunii, "Graph-based design specification of parallel computation," Lecture Notes in Computer Science 163, VLSI engineering, Springer-Verlag, 1984.
(16) H. Kobayashi, N. Endo, T. Nakamura and Y. Shigei, "Performance evaluation of a general purpose pipeline system," Trans. IECE Japan, vol. J68-D, no. 10, pp. 1744-1752, Oct. 1985.

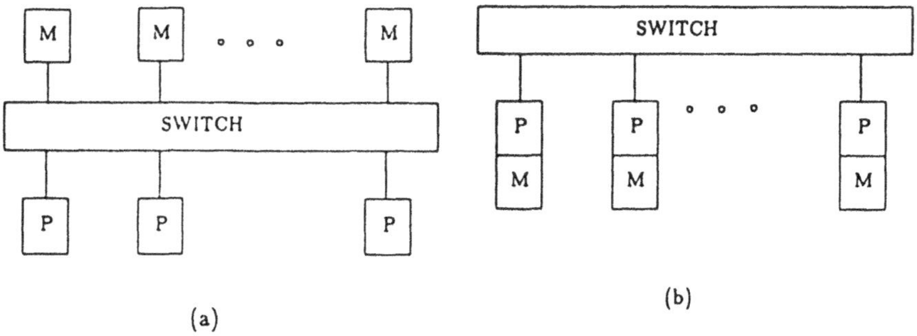

Fig. 1. Multiprocessor model 5. (a) Shared memory. (b) message-passing.

```
c_0 = 0
FOR i FROM 1 to 8 DO
    a_i = d_i / e_i
    b_i = a_i * f_i
    c_i = b_i + c_{i-1}
ENDFOR
```

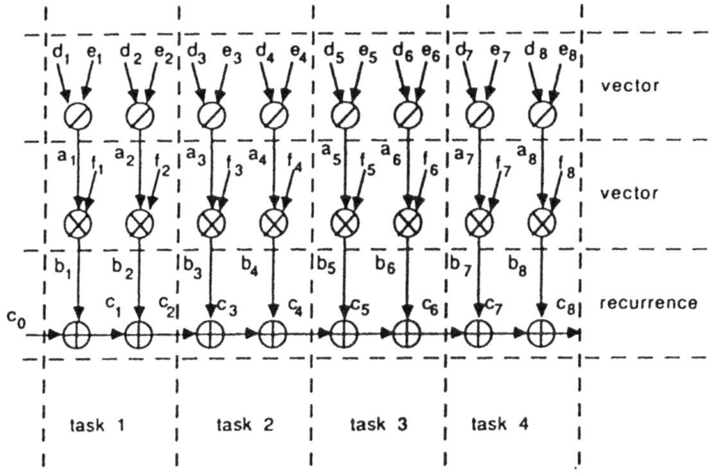

Fig. 2. Program slicing imposed by a vector machine and a multiprocessor.

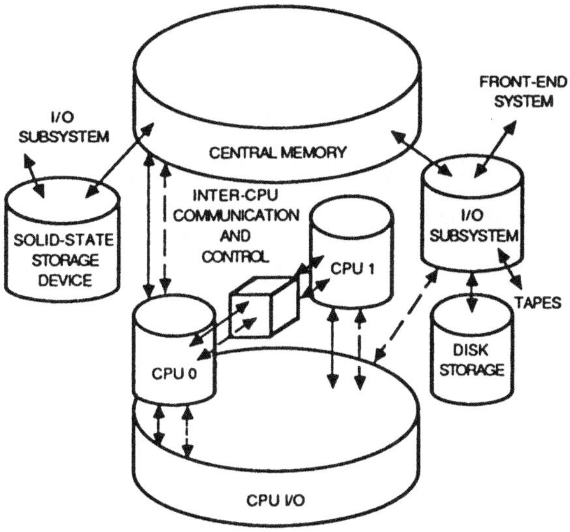

Fig. 3. The system organization of the Cray X-MP-2.

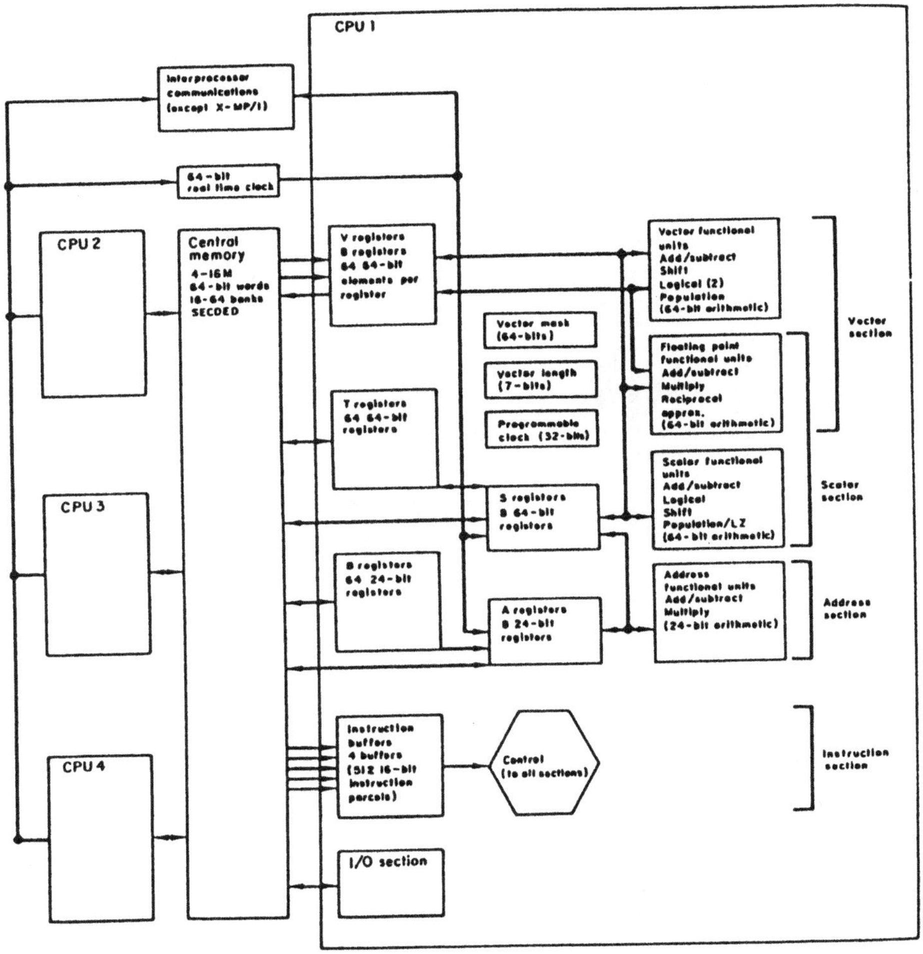

Fig. 4. Cray X-MP-4 architecture.

(a)

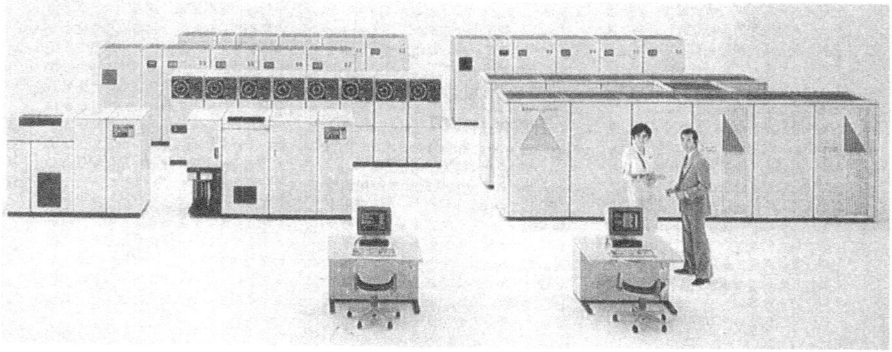

(b)

(c)

Fig. 5. Japans's supercomputers. (a) Facom VP-200 (b) Hitachi's-810. (c) NEC SX-2.

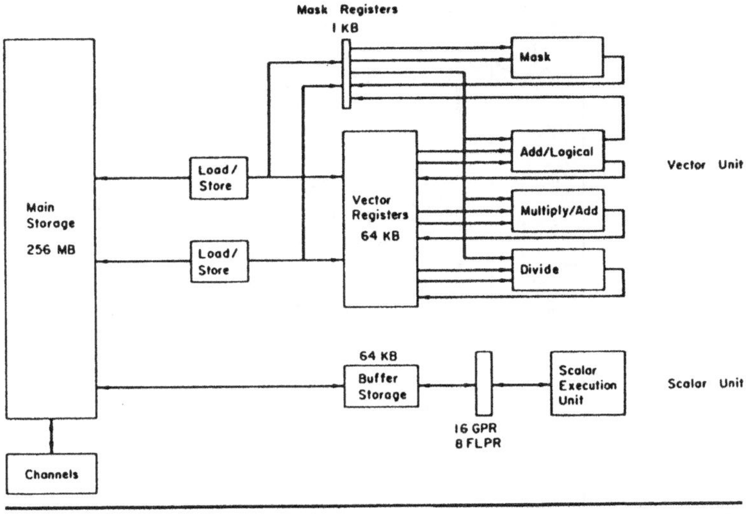

Fig. 6. Fujitsu VP-200 Architecture.

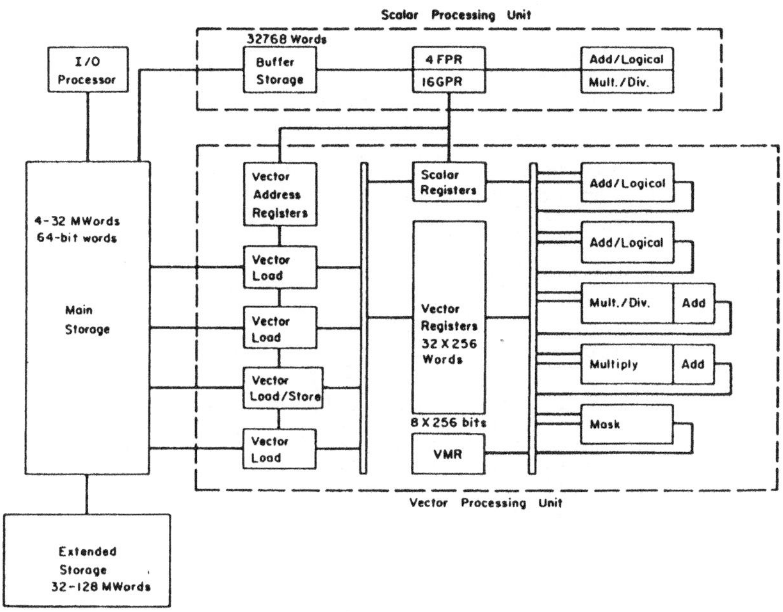

Fig. 7. Hitachi S-810/20 architecture. FPR = floating-point registers; GPR = general-purpose registers; VMR = vector mask registers.

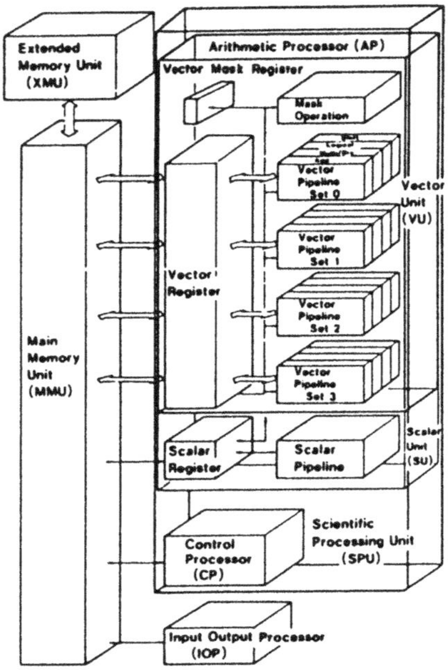

Fig. 8. SX-2 System Configuration.

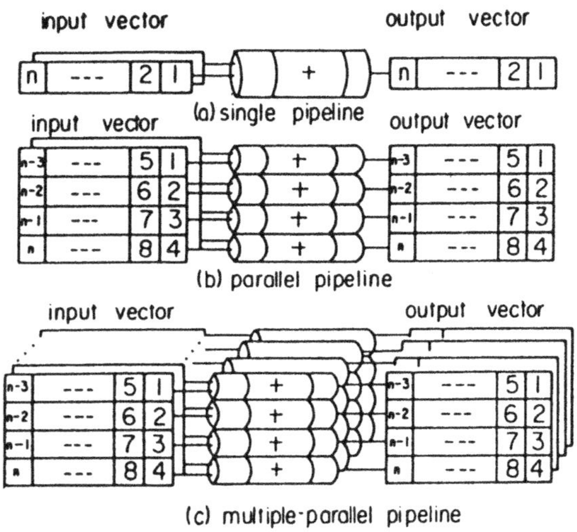

Fig. 9. Various Pipelines

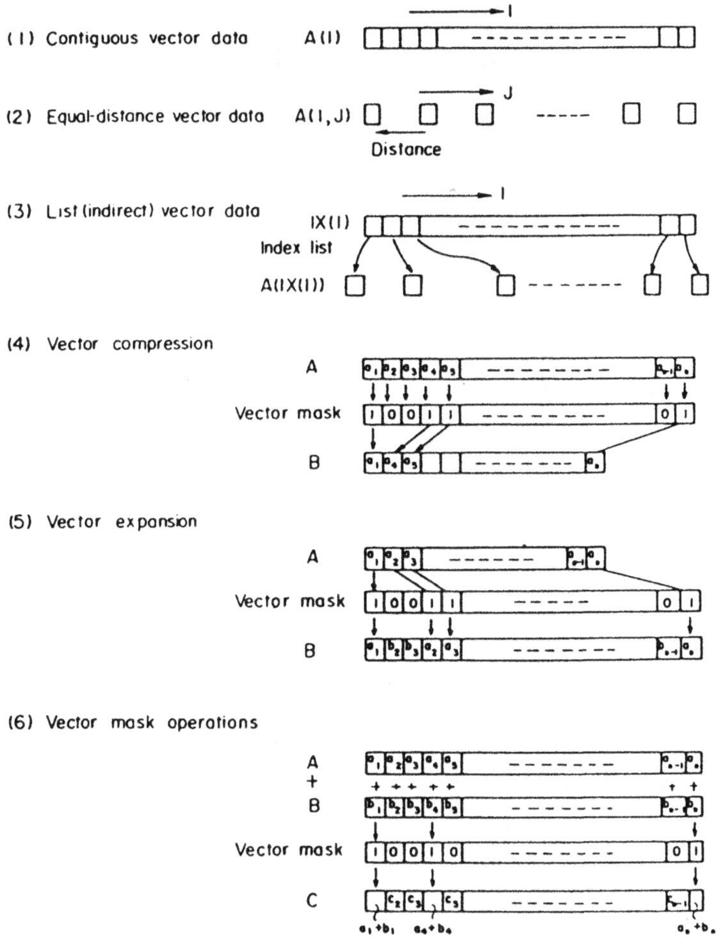

Fig. 10. Various Vector Functions

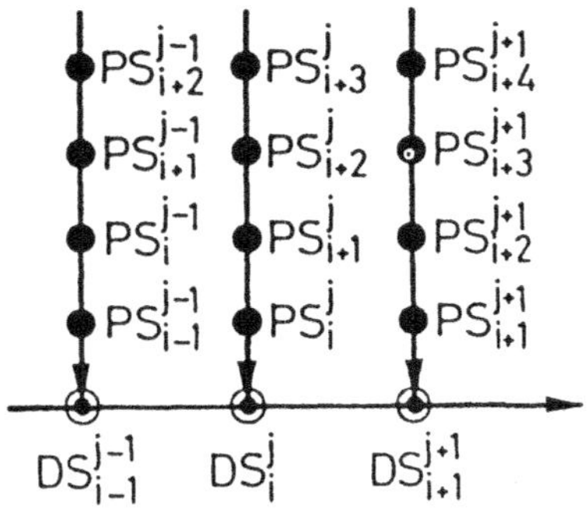

DS_i^j : data of the j-th task within PE_i

PS_i^j : program for processing DS_i^j within PE_i

Fig. 11. Scheme of the pipeline in an MISD fashion.

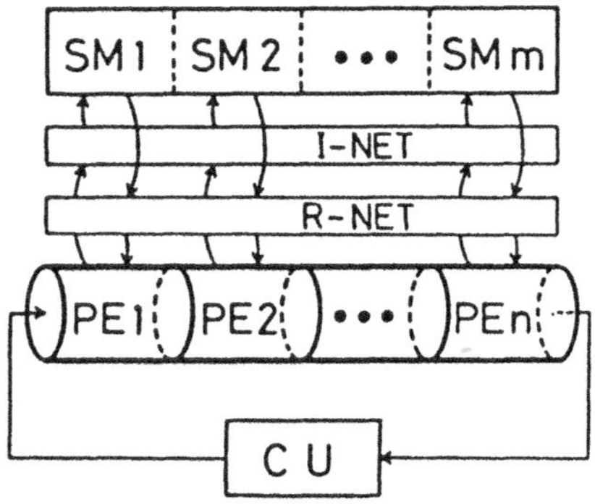

Fig. 12. Overview of general-purpose pipelining based on multi-processing.

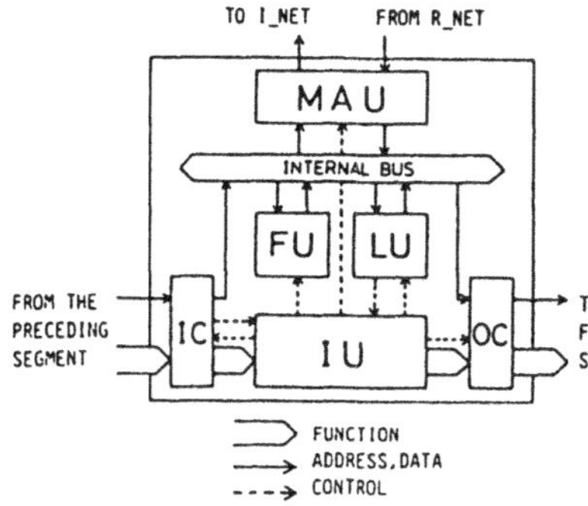

Fig. 13. Processing element.

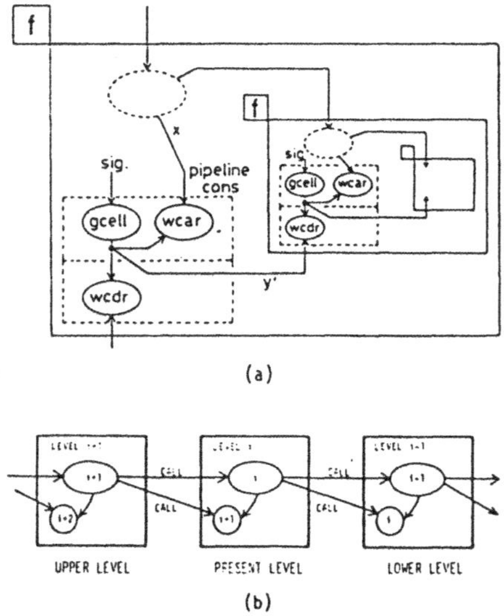

Fig. 14. Scheme of pipeline cons.
(a) Scheme of any instance.
(b) Chining of instances

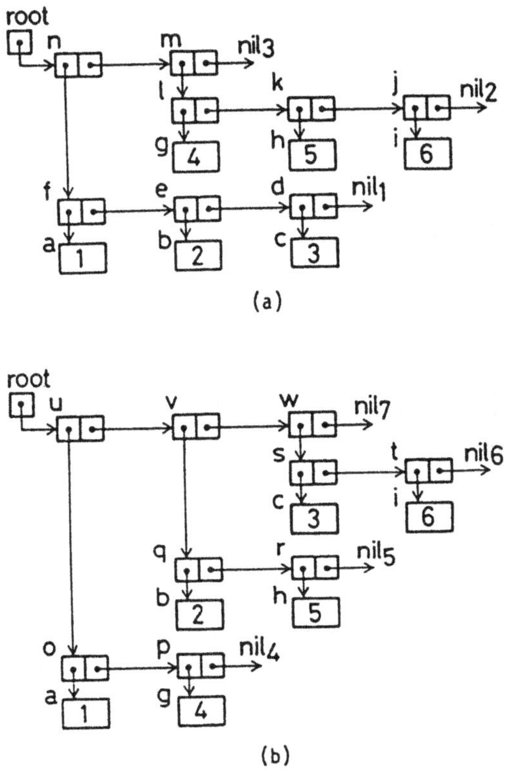

Fig. 15. Object represented in a list-structure.
(a) Initial object. (b) Transposed object.

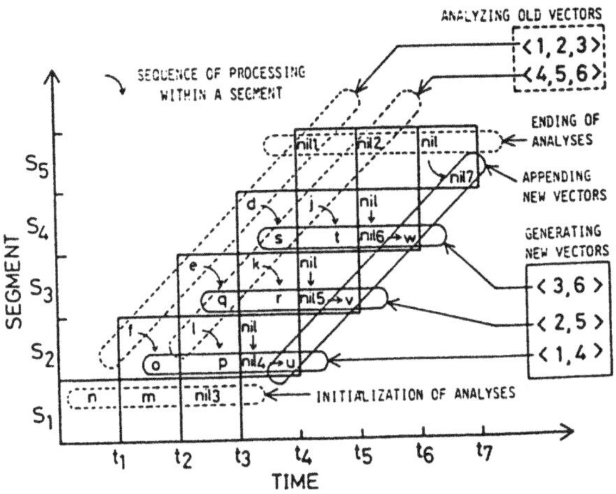

Fig. 16. Pipeline execution of transpose.

FACOM VP2000 Series

K. Uchida

Supercomputer Department, Mainframe Division, Fujitsu Limited 1015, Kamikodanaka Nakahara-ku, Kawasaki 211, Japan

VP2000 series is the supercomputer which attains the maximum performacnce up to 4GFLOPS. With this high performance, VP2000 series is one of the world's highest performance machine as a uni-processor system. In this paper, we would like to introduce the VP2000 series mainly about following items.

-Development of Fujitsu's supercomputer
-Hardware features
-Enhancement of vector and scalar performance
-Dual-scalar processor system
-Enhancement on main-storage and channel-processor
-LSI technology, high-density packaging, cooling system
-Enhancement on software (UNIX system)
-Flexible system

KEY WORD; SUPERCOMPUTER, VECTOR PROCESSOR, VP2000 SERIES, DUAL SCALAR PROCESSOR

1. Introduction;

VP2000 series supports full compatibility between previous VP and VP-E series, and ranked as the new-series with higher performance. VP2000 series consists of a total of 8 models; 4 uni-processor models (VP2600/10, VP2400/10, VP2200/10, VP2100/10) and 4 dual-scalar-processor models(VP2600/20, VP2400/20, VP2200/20, VP2100/20), all of which are field-upgradable within the series. VP2600 offers maximum performance of 4GFLOPS, one of the highest as a uni-processor system. High performanced scalar-unit is implemented, and by connecting 2 of these scalar-units, VP2000 series has attained high speed performance in total.

Fig. 1 FACOM VP2600/20

2. Supercomputer of Fujitsu;

To meet the demands of high speed processing in research, development and engineering area in the age of high-technology, introduction of supercomputer in Japan advanced rapidly. We developed the VP series, taking advantage of studies and experiences gained in

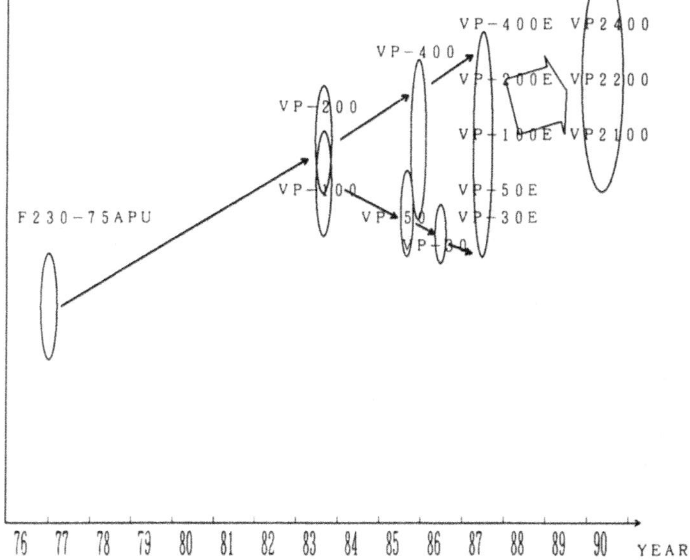

Fig. 2 Fujitsu Supercomputers

F230-75APU system(Developed 1977) which was the Vector processor using pipeline method. After announcing VP-200/100 as the Japan's first supercomputer in July 1982, we announced high-end model VP-400 and low price model VP-50 in April 1985. Classifying the recieved orders by industry, the manufacturing industry covers over 50%. Electronic manufacturing field covers 15% and the rest is covered by chemestry, automobile, and etc.

In the past, there were two demands for supercomputing market. One was for high speed processing and the other was for low price machine, which VP-E series respond. The VP2000 series follows as the new generation processor with much higher performance.

3. Features of VP2000 series;

The main feature on VP2000 series is the upward compatibility to previous VP and VP-E series, and the enhancement in both scalar and vector performance. Employing the leading technology with new designing tools and architectures, VP2000 series seeks high performance, expandabilites, flexibilites, open system, and smaller operation costs.

The new features supported in VP2000 series are the adoption of dual scalar processor, the expansion of main storage, the introduction of system storage, and the implementation of UNIX environments.

4. Hardware of VP2000 series;

4.1 The models;

The VP2000 series covers a wide vector performance range, with the most powerful model being 8 times faster than the low-end model. Dual scalar processor system is adoptable to improve scalar performance and to increase the efficient usage of Vector unit. All models are field upgradable within the series.

VP2000 series supports full upward compatibility with current VP and VP-E series, that no conversion is necessary on software programs.

4.2 The hardware;

The Fig.4 shows the hardware organization of VP2000 series. The basic hardware units of the VP2000 series are the vector processing unit, the main storage unit, the system storage unit, and the channel unit. The vector processing unit consists of vector and scalar unit. The vector unit has vector registers, mask registers, and vector pipelines each performing high speed operations. The scalar unit attains high performance with powerful execution units, improved scalar pipeline(1-cycle pipeline), and large buffer storage.

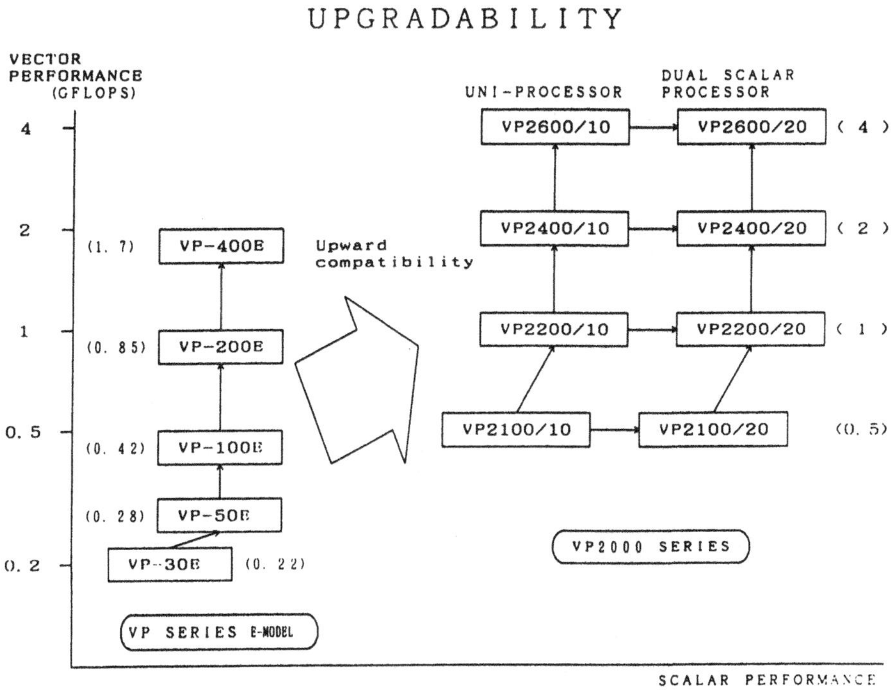

Fig.3 Updradability

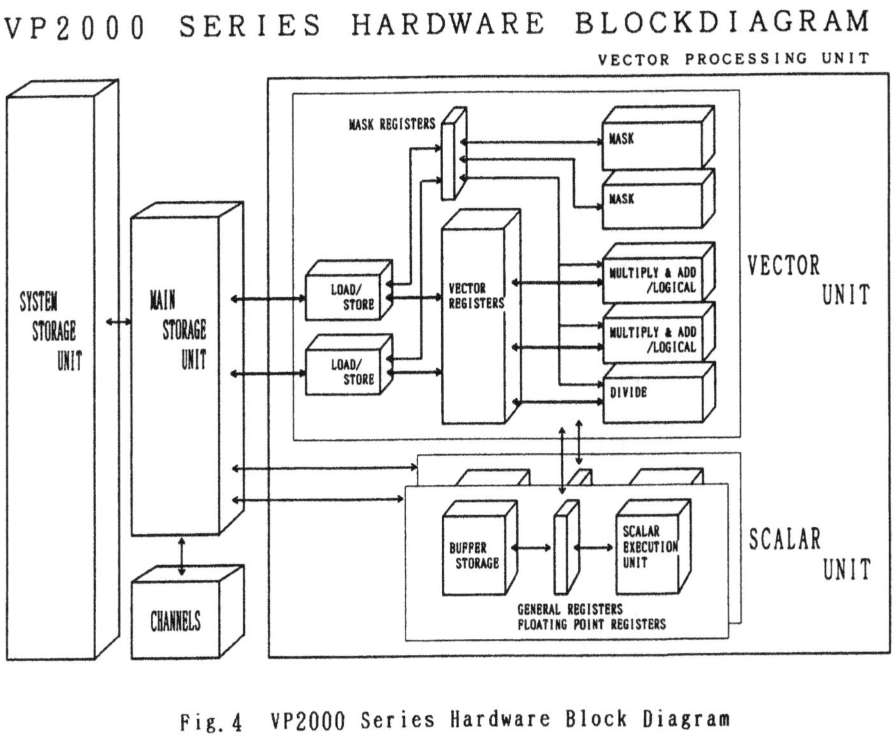

Fig.4 VP2000 Series Hardware Block Diagram

4.3 Specification;

The top model VP2600 has the total Vecter Register capacity of 128KB(64word x 8byte x 256). To make the best utilization of the total capacity, for example, the Vecter Registers may be concatinated to take the following configurations: 64(length) x 256(total number), 128 x 128, 256 x 64, ..., 2048 x 8. The length of Vecter Register is specified by a Vector Control Register, and can be altered by an instruction in the program. Mask Register has total capacity of 2KB and has the same configuration as Vector Register. Mask Register can also be concatinated just as same as Vector Register.

VP2000 SERIES SPECIFICATIONS

	VP2600/20 VP2600/10	VP2400/20 VP2400/10	VP2200/20 VP2200/10	VP2100/20 VP2100/10
VECTOR PEAK PERFORMANCE	4 GFLOPS	2 GFLOPS	1 GFLOPS	0.5 GFLOPS
NUMBER OF VECTOR UNIT	1			
NUMBER OF SCALAR UNIT	1 ~ 2			
CAPACITY OF VECTOR REGISTER	128KB/SU	64KB/SU	32KB/SU	
CAPACITY OF MASK REGISTER	2048B/SU	1024B/SU	512KB/SU	
NUMBER OF VECTOR PIPELINE	7			5
VECTOR PIPELINE THROUHPUT	4	2	1	
MAIN STORAGE CAPACITY	128MB ~ 2GB	128MB ~ 1GB	64MB ~ 1GB	32MB ~ 1GB
SYSTEM STORAGE CAPACITY	1, 2, 4, 6, 8 GB			
CHANNE — NO. OF CHANNELS	16~128			
CHANNE — THROUGHPUT	Max. 1GB/S (OPTICAL CHANNEL:9MB/S, BMC:4.5MB/S)			
CAPACITY OF BUFFER STORAGE	128KB/SU			

VU:Vector Unit, SU:Scalar Unit

Fig.5 VP2000 Series Specifications

4.4 Achievement of High Performance;

The following architecture improves performance of the VP2000 series.

4.4.1 Enhancement on Vector Pipelines;

In VP2000 series, identical 2 execution pipelines are implemented naming the "Universal Pipeline". Each Universal Pipeline can execute Multiply, Add, Multiply & Add and Logical Operation. By operating these two in parallel, execution performance is higher than that of previous VP series. The performance in Divide Pipeline is also enhanced in the new series.

4.4.2 The enhancement on scalar performance;

One-cycle pipeline is implemented in scalar unit in order to enhance the scalar performance by decreasing instruction execution cycles. The overlapping of execution sequence enables to issuse the next instruction one cycle after the issue of previous instruction which contributes to higher performance. The Buffer Storage is also extended to 128KB in VP2000 series.

4.4.3 Dual Scalar Processor;

The Dual Scalar Processor is a newly developed architecture to increase the efficiency of vector unit, shared by two scalar processors. By implementing this architecture, the performance of a single job is raized up to 2 times, and the maximum throughput 1.5 times that of usual multi-processer system, so that price/performance can be significantly improved.

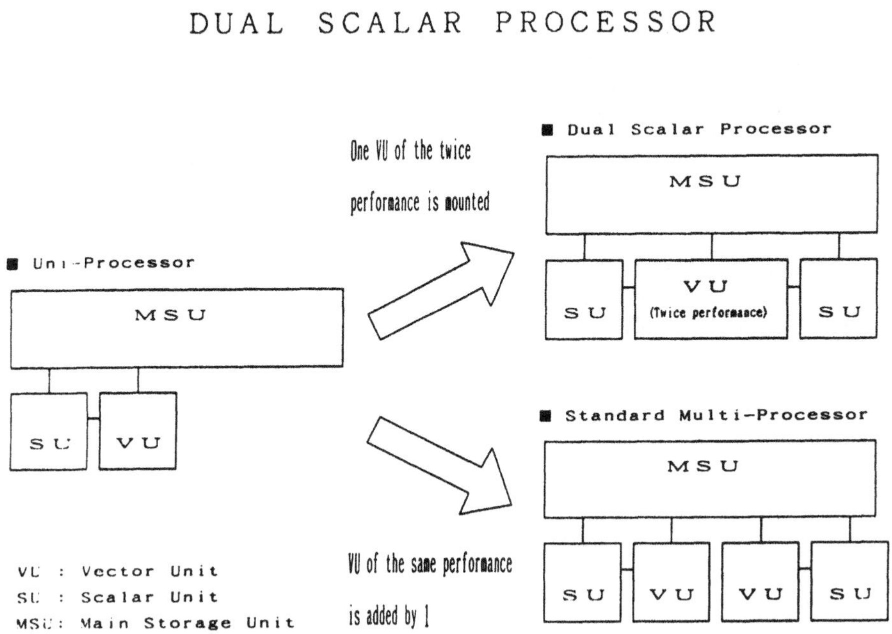

Fig.6 Dual Scalar Processor.

4.4.4 Enhancement of Memory System;

High access speed and throughput is realized by extending the maximum storage capacity up to 2GB, using high speed 1Mbit static RAM. 8GB maximum System Storage is also supported in order to raise the system throughput by utilizing to the high speed Input/Output mechanism and to the swapping area of vector jobs. Thus the flexible utilization environment is made possible for example, to increase the TSS-terminals to use the vector processor directly or to execute large back-ground jobs in ordinal utilization time.

4.4.5 High throughput channels;

High throughput of 1GB per second is attained in maximum configuration of 128 channels. Both normal channels(4.5MB per second) and Optical channels(9MB per second) can be connected in VP2000 series. Optical channels can be connected to peripheral devices by using optical fiber cable, and can be placed as far as 2km distance.

5. Advanced Technology;

New technologies were developed for VP2000 series. VP2000 series requires high performance in technology. High-speed -LSI's and static RAM for the main storage, high-density printed board equipped with many input/output pins, and high-efficiency conduction cooling modules are implemented for VP2000 series.

5.1 Advanced LSIs;

The logic circuits are designed using high-speed, high-density emitter-coupled-logic (ECL). Each ECL LSI can pack 15000 of 80 ps delay gates. In addition, 64-Kbit Random access memory with 1.6 nano seconds access time are packed with 3500 logic gates together on the same chip. For the main storage, 1M-bit large-capacity static random access memory chips with 35 nano seconds access time were adopted.

Fig.7 LSIs

5.2 High Density Packaging;

Fujitsu is the first to have introduced glass ceramics as a board material for greater transmission speed. Each vecter unit or scalar unit is made up of boards with high-density and multi-layer integration of a maximum of 144 mounted LSI's on a 24.5cm square ,61-layer glass ceramics board.

Fig.8 High-Density Multi-layer Grass Ceramic Board and Conductive Cooling Module.

5.3 Cooling Technology;

To deal with the large amount of heat generated by super high speed and high-density LSIs, a new high performance conductive cooling module(CCM) is introduced. CCM has a superior cooling capacity and permits a max heat dissipation of 30W per LSI, and 4.6KW per board. By hermetically sealing the connections between CCMs and MLG, interface is protected from humidity and increases the reliability.

6. Software of the VP2000 series;

In recent years the environment for supercomputer has extended and more-easy-to-use open system is required, such as the utilization of UNIX and Workstation which spreads throughout the research and development area. Especialy, the performance of workstation is improving exceedingly, so the requirement for distribution of functions and the cooperations between host·machine and workstation is raising. Since UNIX is the mainstream of operating system for workstation now, VP2000 series is enhanced also in software to meet the needs in such environments.

6.1 Vector Processing Support on UNIX ;

Vector processing supports "UTS/M + VPO" as a control program of the VP2000 series and a vectorized FORTRAN compiler has been added to UTS/M. Using this software, all kinds of work can be performed in the UNIX environment, from efficient program development utilizing UNIX features to program executions of vector processing using FORTRAN language. UTS/M is based on UNIX System V, an operating system developed by AT&T, and includes the BSD functions such as TCP/IP, the functions necessary to support the host machine, and independent functions such as extended addressing feature. Furhtermore, X-Windows and NFS (Network File System) are supported, so that vector processing on the CPU of the VP2000 through the multi-windows, and local processing on workstations can be processed concurrently, and it is feasible to use files on the host machine directly from workstations. Then, it is available to use the UNIX environment consistently during program development and vector executions.

SYSTEM STRUCTURE OF UTS/M

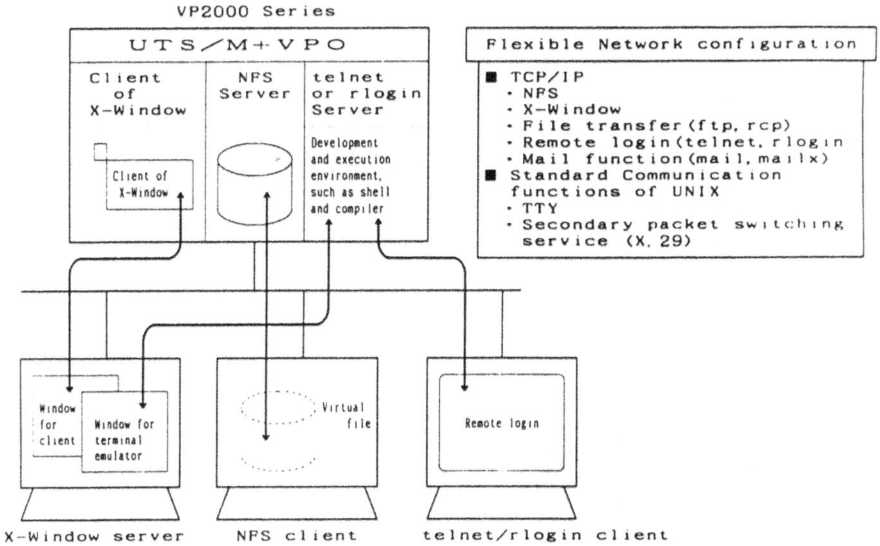

Fig. 9 System Structure of UTS/M

6.2 TCP/IP support on the MSP;

TP/IP is supported on the Fujitsu's general purpose operating system(MSP), to connect UNIX workstations. Supporting TCP/IP directly on MSP, VP2000 series can be connected with workstations without any supplement on hardware nor software. Moreover supporting TCP/IP, following operations are also feasible:

- SMTP(Simple Mail Transfer Protocal), the communication between mail functions in UNIX environment and DSM by FUJITSU.
- File transfer between VP2000 series and workstations with CP/IP protocol.
- Utilization of TSS from workstations.

6.3 Flexible system configuration;

In VP2000 series, various system configurations, OSs and environments are offered to meet the varying needs of customers.

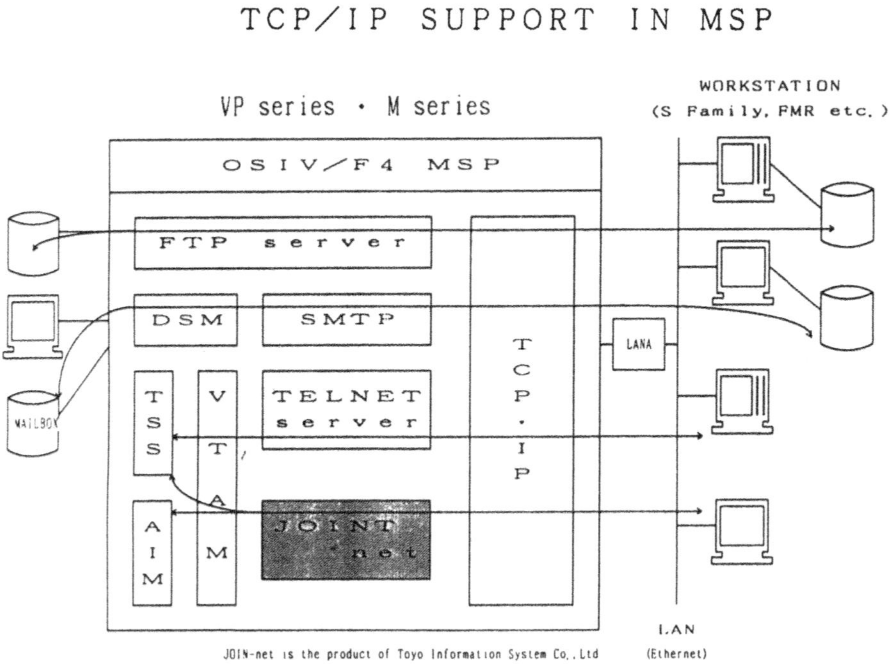

Fig. 10 TCP/IP support in MSP

6.3.1 Flexible system configuration;

Two system configurations are available in VP2000 series, which is all the same to the previous VP series. One is the "loosely coupled backend system" which we can use VP with maximum preformance through a Front End Processor(FEP). Another is a "stand alone system", where a VP performs all the processing.

6.3.2 Multiple OSs under AVM(Advanced Virtual Machine);

Multiple OSs can run simultaniously on the VP2000 series as same as previous VP series. For example, UTS/M and MSP. It can also be used as a means of migration tool between OSs. In addition, for VP2000 series, function of communication between UTS and MSP on AVM is enhanced. It is now possible for a UTS user to access MSP facilities, to transfer files between main storage and other devices in higher speeds. It is also possible to schedule jobs automaticlly using HICS(Hierachical Information Control System). For instance, it is possible to start a job after transfering the files by HICS from other machine.

6.3.3 Processor configuration of DSP;

The dual scalar processor (Fig. 11) has symmetric and asymmetric configurations. The former shares one vector unit so that the vector unit is active at the maximun efficiency. For this reason high system throughput for an ordinary user is obtained, whose jobs are a mixture of vector and scalar jobs. In the latter, one scalar unit connects with a vector unit and the other scalar unit without a vector unit. An asymmetric system is suited for users such as laboratories whose jobs include high vectorization ratio and large vector jobs.

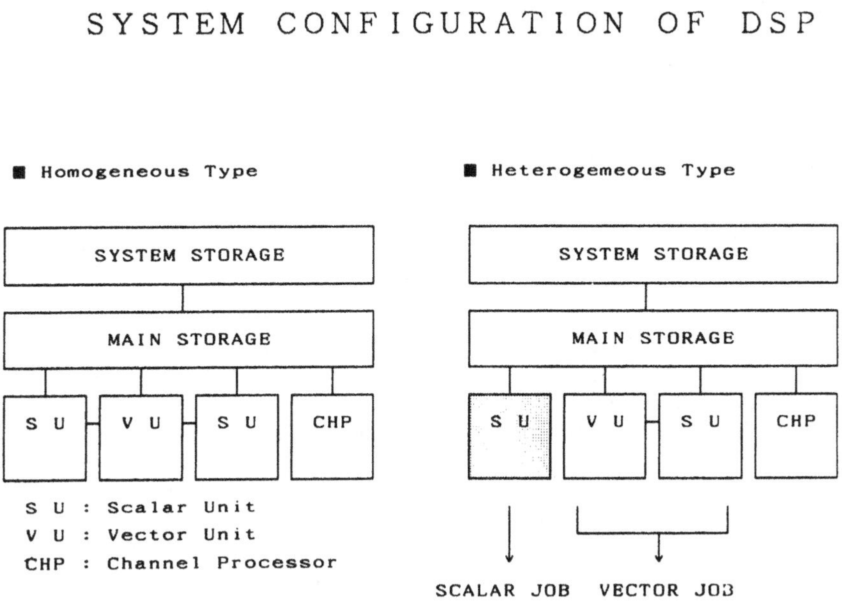

Fig. 11 System Configuration of DSP

7. conclusion;

This paper described the introduction of FACOM VP2000 series supercomputer. As explained, the FACOM VP2000 series has new features in architectures, LSI technologies, and softwares. There are infinite demand for higher performance in area where supercomputer are utilized. In order to meet the demand, Fujitsu will continue to pursue the development of high speed processing, parallel processing technology, new architecture, and useful software.

Expert Systems Based on Object Model - An Approach to Deep Knowledge Systems

H. Ueno and Y. Oomori

Department of Systems Engineering,
Tokyo Denki University, Saitama 350 - 03, Japan

ABSTRACT

This paper discusses the need for deep knowledge systems, the concept of object model, a systematic modeling method and applications. The object model is useful to represent the knowledge about structures and behaviors which are known as one of deep knowledge for representing complex structured objects. Rule-based systems, i.e., surface knowledge systems, are convenient to represent expertise, but their performance and flexibility are limited. Advanced knowledge-based expert systems to solve complex problems need to handle both types of knowledge. A way of combining rule-based reasoning and model-based reasoning is shown. The object model can be represented in a extended frame-based knowledge representation formalism by means of a combination of ISA and PART-OF relations. Knowledge engineering environment ZERO has been developed based on this concept and has been used for developing a variety of problem solving systems. Three applications are shown from different domains which are trouble shooting expert systems for electro-mechanical devices and for petroleum refinery plants and a knowledge-based program understander as an aid for novice programmers. The most important feature of the deep knowledge systems is that the knowledge should be shared by a variety of applications. The object model based knowledge representation schema may meet this requirement.

INTRODUCTION

From experiences over years of developing expert systems we have learned that in order to achieve high performance and flexibility in problem solving, features to represent and manipulate deep knowledge are the key issues. This is understandable because human experts have the knowledge about basic theories and principles extracted from education and research, in addition to expertise obtained through a long time experience in a specific domain. In another case we know that design sheets and design specifications are very useful for engineers who are trying to find failures within systems, or are trying to understand texts describing complex devices. It should be noted that these types of knowledge which are referred to as "deep knowledge" are common and domain independent, and can be shared by a variety of problem solving methods. The knowledge about principles and theories and the knowledge about structures and behaviors are known as typical deep knowledge, e.g. Davis[9], Ueno[12]. For this reason we can suppose that knowledge-based systems having features to manipulate these kinds of knowledge may be applicable to a variety of domain problems.

Problems to be solved by knowledge-based approaches are divided into two contrastive groups which are analysis-oriented problems such as medical diagnosis and trouble shooting, and synthesis-oriented problems such as intelligent CAD, scheduling and design, e.g. Davis[9], Hart[10], Ueno[7,13]. The typical problem solving strategy in the analysis-oriented problems is to choose the most likely hypothesis from a set of hypotheses by analysis of given data using the knowledge stored in the knowledge base. On the other hand, the typical problem solving approach in the synthesis-oriented problems is to construct the most appropriate object by combining suitable elements under the given constraints to meet the given requirements. The deep knowledge systems are expected to solve both kinds of problems just like human experts using the same knowledge.

An important feature of deep knowledge should be universality in interpretation which results in high flexibility and performance in problem solving, while surface knowledge which has been applied to most currently available expert systems is dedicated to specific interpretation in specific problem solving within a narrow domain, and therefore resulting in limited flexibility and performance. This means that appropriate modeling methodologies are required to achieve knowledge independence.

In this paper we deal with the second type of deep knowledge, i.e., the knowledge about structures and behaviors, which are useful to manipulate complex structured systems. The object model-based knowledge representation schema are considered to be shared by a variety of applications which include both analysis- and synthesis-oriented problems for complex structured systems. We call such a representation model an "object model".

From the point of view of computer processable knowledge representations structured systems include both abstract-specific relationships and whole-part relationships in nature, e.g. Ueno[7]. The first one is used to represent an element hierarchy, and is naturally represented by frame model, e.g. Minsky[3] since the inheritance control function for attribute information is naturally applicable in this hierarchy. The second one is used to represent the structure of the system in combination with a semantic network representation formalism, and is not appropriately represented in frames, because inheritance is not applicable to this relationship in a hierarchy. Therefore we need a powerful and flexible knowledge representation framework for the object model. We have solved this problem by means of an extended frame formalism by Ueno[7], where the inheritance is extended to apply to a set of slots, i.e., elements of frames, so that the knowledge about structures and behaviors are inherited as a set. In an ordinary frame system, inheritance is controlled on a slot-by-slot basis. By this extension, template schema of structured objects are inherited naturally to their subclasses and instances.

In this paper we show the concept of object model, the modeling method and examples of applications. The application systems are implemented in frame-based knowledge engineering environment ZERO, by Ito and Ueno[6,11], which is extended to handle the extended frame representations for the object model-based system, e.g. Oomori and Ueno[19]. The extension was accomplished so that the concept and methods of the object model can be realized smoothly. The ZERO system is written in the COMMON LISP language and is available on a variety of computer systems, from FACOM main frame computer systems to microcomputers. ZERO is used to develop a variety of knowledge-based systems such as trouble shooting systems, natural language interface systems, intelligent CAD systems, intelligent programming environments and financial advice systems.

CONCEPT OF OBJECT MODEL

By "object" we mean the objective of problem solving by means of knowledge-based approaches. It includes a physical object, a conceptual object and a system in general. We call a representation model of such objects an "object model", by Ueno[13]. For example, the computer model of a computer is the object model when the computer is an objective of knowledge-based problem solving. In case of an intelligent program development support system an objective is a computer program in terms of intelligent handling, and an abstract representation model of the program is the object model. The object model could be represented by means of symbolic methods or analogue expressions. These are determined according to the the characteristics of objects and their problem solving. This paper does not discuss all of them but those aspects of a system which has physical or logical structure. In any way the object model is one of the representation models for deep knowledge as well as for semantics. We believe that features to manipulate the object model will become very important for advanced knowledge-based expert systems in the future.

Problems to be solved by knowledge-based approaches are divided into two groups; synthesis-oriented problems and analysis-oriented problems. Constructing the model of an object by assembling available parts under the specific constraints to meet a set of given requirements should be the issue in the synthesis-oriented problem. For analysis-oriented problems also, the object model plays important roles. We know this from the following example. A person who has the knowledge about the principle and the structure of one thing can solve a variety of problems asout it using the knowledge. For example, a person who knows the principle and the structure of a computer system can express what a computer is to another person, can understand a picture of a computer, can find some failures in the design specifications of a computer, and can do some trouble shootings for a computer. Of course, the quality and the performance of these abilities depend on the quality and quantity of the knowledge he has about a computer.

On the other hand, currently available knowledge-based expert system based on surface knowledge such as a Production System, where a unit of knowledge is represented in an IF-THEN rule, can work only for solving specific problem on specific aspects in a narrow domain, e.g. Davis[9], Hart[10]. This is because these expert systems have limited flexibility and therefore the ability of the system is much inferior to that of human beings. From the point of view of problem solving method, it seems unlikely that a human being treats an analysis-oriented problem and a synthesis-oriented problem as different ones. Instead, it seems true that he can use common knowledge for a variety of problems in various ways according to situations he is faced with.

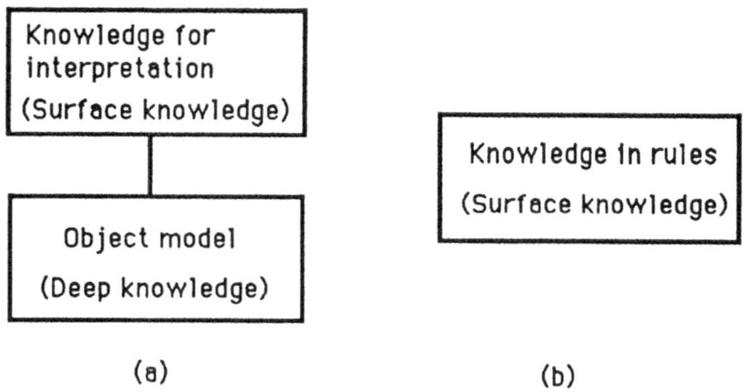

Fig.1 Two-layer knowledge base based on the concept of the object model(a), and a traditional single layer knowledge base in a surface knowledge representation(b).

The structure of a knowledge base to meet these circumstances should have a two-layer hierarchy shown as in Fig.1a where the lower layer consists of the object model ,i.e., deep knowledge, and the upper layer surface knowledge ,i.e., a set of production rules. A usual knowledge base has a flat structure as shown in Fig.1b. It must be noted that a three-layer hierarchy model should be suitable for certain applications. It is not an issue whether we must choose two rather than three. The issue is that a multi-layer hierarchy model should be applied to organize a knowledge base in order to achieve highly flexible and powerful expert systems.

What kind of data structures is the most suitable to represent the object model? It should be emphasized that there is no single and universal structure. Instead, the suitable data structure should be chosen according to the type of the problem to be solved. For example, for a mechanical CAD system the object model has to be represented in a combination of structures, behaviors and other attributes since the object to be dealt with has a three dimensional structure and specific functions. For a dynamic analogue system the object model should be a set of simultaneous differential equations. For medical diagnosis based on pathophysiology the object model should be represented by means of a transition network based on a concept of pathophysiology. As shown here a model description seems to be dependent upon relevant problem domains. From the point of view of representing the model, however, a semantic network representation model seems to be widely applicable as a common representation model. In addition, a large complex object should be represented in a hierarchy, because the number of similar modules(parts) must be included in the model.

As a conclusion we think that the object model can be appropriately represented in an extended frame-based representation formalism to construct a knowledge base. Because the frame model has flexible and powerful functions which comes from useful features such as hierarchical data structures with attribute inheritance functions, a combination of declarative and procedural knowledges, allowing a variety of data types and a suitable extension should meet our requirements.

In the following sections we describe how the object model can be represented in the extended frame-based formalism. We propose basic modeling rules and a systematic procedure for model description for complex structured object by the extension. This modeling method may be quite useful not only for a deep knowledge-based diagnosis system but also for an intelligent design assistant system, which means that the methodology of the object model seems to be able to bridge more easily between analysis-oriented problem solving systems and synthesis-oriented problem solving systems.

REPRESENTATION OF OBJECT MODEL IN FRAMES

We show how the object model can be represented in the extended frame-based formalism using a very simple example from a block world. At first ordinary frame-based representations are shown. Then the need for extensions is pointed out, followed by the extended frame-based modeling.

Fig.2 shows a frame-based representation of an arch which is constructed with three blocks, where the structure of the object is represented by means of a combination of PART-OF relations, i.e., whole-part relations, and semantic networks in the left hand side of the figure and the elements are represented by ISA relations, i.e., abstract-specific relations, in the right respectively. This figure means that a structured object has two different aspects. We should combine these representations as single model according to the concept of the object model.

Fig.3 shows a method of representing the object by simply connecting the associated nodes between both representations. However, this method does not fit in the frame-based formalism because an inheritance control function can only be applicable to the ISA relations by nature. We need a suitable modeling method so that the inheritance control function can be applied to the structure representations as well as the element representations, because a structure is a kind of object attribute and a structure description of the class object should be inherited to the instance representation of the objects. Otherwise, model descriptions and manipulations will be too complicated. Therefore, this type of ordinary frame-based modeling is not suitable to represent large complex systems and an alternative method is then required.

New modeling method, which we call as the extended frame-based formalism, is shown in Fig.4, where an arrow shows a relationship between a root node and a destination node. For example, the relation "an arch is a (kind of) a physical object" is written as "Arch ISA PhysicalObject" and an ISA arrow is drawn from Arch to PhysicalObject. For another example, the relation "an arch has part A" is written as "Arch HASPARTS A" and a HASPARTS arrow is drawn from Arch to A. This is also rewritten as "A is a part of an arch" and as "A PARTOF Arch", then this relation can be written by a PARTOF arrow from A to Arch.

The modeling concept is as follows. An arch itself is considered to be the principal element in representing an arch system. Therefore the

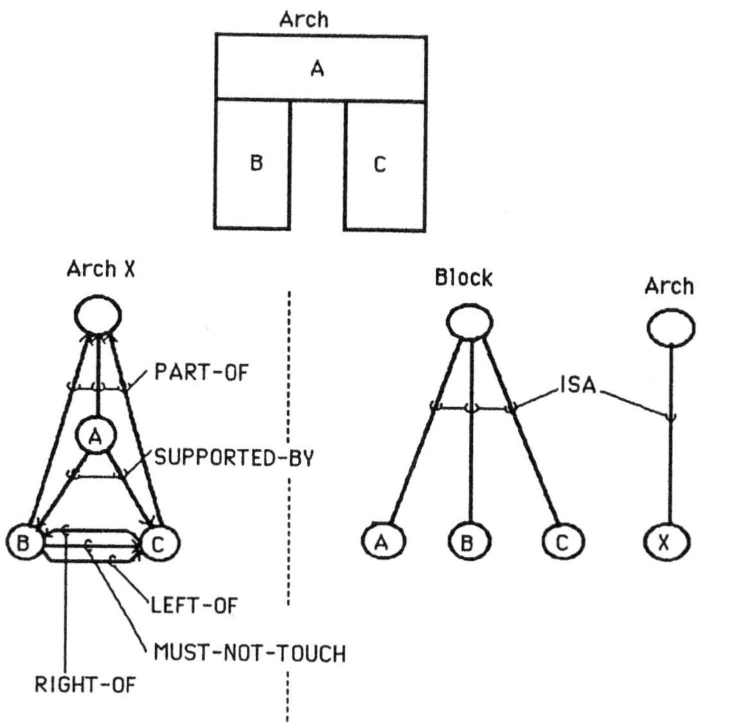

Fig.2 A frame-based representation of an arch. Structure description(left) and component description(right).

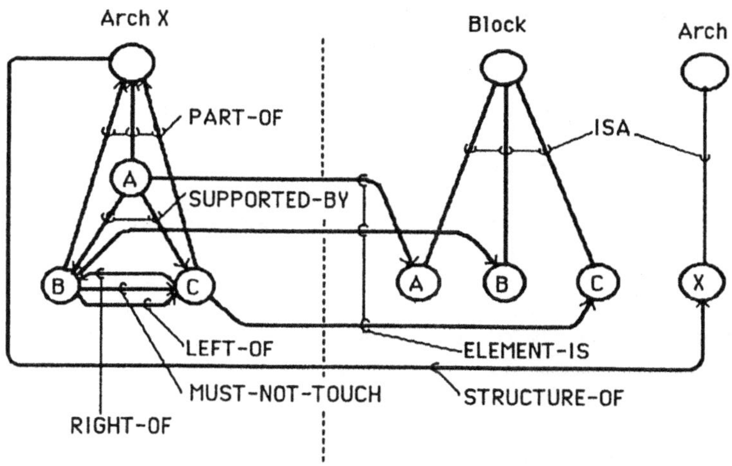

Fig.3 A frame-based representation of an arch by connecting a structure description and component descriptions.

representation should be done around the arch element and the rest of the elements should be represented as components of it. Now, the model representation according to the concept is shown as in Fig.4, which is designed in the following manner. First, since an arch is a physical object and arch X is a kind of an arch, i.e., an instance of the arch class, the relations among these three objects are represented as in the left by two ISA relations in a hierarchy (for more precise representation the lower relation should be drawn by an INSTANCE-OF relation). In the right the structure of the arch is represented by linking their parts to each other as follows: From the top to the bottom there are ISA relations such as "a block is a physical object", "a brick is a block", "block A is a brick", "block a is an instance of A", and so on. In terms of a structural representation, it must be stressed again that the structure of the arch should be defined at a prototype level as a class object, i.e., the level where the arch node is defined, so that it can be copied to the instance level, because the structure is the same between these two levels as in the figure. HASPARTS, RIGHT-OF, SUPPORT and MUSTNOTTOUCH relations play important roles in structural representations in addition to the ISA relation.

By such structural representations it becomes possible to extend the usage of ISA relations originally used in element-by-element basis in a hierarchy. In order to achieve the extension all components connected by HASPARTS links must be included within a single frame as a set of attributes of the frame. That is, the structure descriptions of a class frame "arch" can be inherited to the instance frame "arch X" as a set of attribute descriptions. The extension of the inheritance can be naturally achieved in extended frame-based knowledge representation formalisms in ZERO as in the following.

The "arch" and its instance "arch X" can be represented in ZERO as shown in Fig.5 in the following way. Each frame has a unique frame name and consists of a set of slots as components which store attribute data associated to the frame. Each slot has a unique slot name within a frame and consists of five items, i.e., a slot name, a datatype specification, slot value(s), demon(s), and an option. In addition to three system slots, i.e., the slots installed in the system, which are ISA, HASPARTS and RELATIONS slots users can describe a variety of representations by means of user slots, i.e., slots defined by users

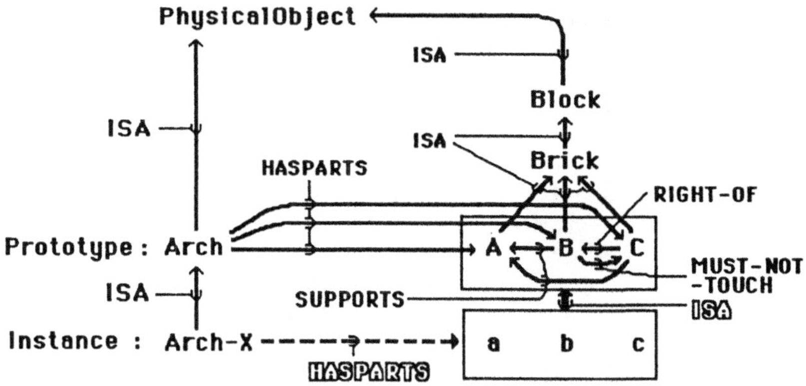

Fig.4 A new modeling method of representing an arch in an extended frame representation scheme.

FRAME: Arch		
ISA	FRAME	PhysicalObject
HASPARTS	FLIST	A, B, C
RELATIONS	PREDICATES	(SUPPORT B A), (SUPPORT C A), (RIGHTOF C B), (MUSTNOTTOUCH B C)
HEIGHT	VALUE	NIL(Default 10)
LENGTH	VALUE	NIL(Default 10)
WIDTH	VALUE	NIL(Default 5)
FUNCTION	TEXT	Toy

FRAME:	Arch-X	
ISA	FRAME	Arch
HASPARTS	FLIST	(a b c)

Fig.5 An example of frame descriptions for Arch and Arch-X according to the concept of the object model.

themselves. Data type specifications used in the following examples and their meanings are:

FRAME:	the slot value is a frame name, i.e., a pointer to another frame,
FLIST:	a list of frame names,
PREDICATE:	a set of predicates,
VALUE:	a numerical value,
TEXT:	a text,
LISP:	a Lisp function name, i.e., an attached procedure's name.

An attached procedure is a procedural knowledge associated to a specific frame, as well as a demon.

Fig.5 shows an example of a frame description of the prototype frame "arch" and its instance "arch X" in Fig.4 in the ZERO formalism. Each slot has a specific meaning as follows. The value "physical-object" of the ISA slot of the "Arch" frame means that the upper concept is a physical object. The list (A B C) which is the value of the HASPARTS slot indicates that the arch has A, B and C as structural components. The attribute data for these components are stored in frames "A", "B" and "C" respectively. The RELATIONS slot has the relationship attribute among the components which is defined by a set of predicates such as "(SUPPORT B A)". The HEIGHT, LENGTH and WIDTH slots contain data about the size attributes of the Arch object. (Default 10) means that the default value for HEIGHT is 10, i.e., if a specific information will not be assigned to the HEIGHT slot of the instance of this class frame 10 is assigned instead. The value "toy" of the FUNCTION slot means that the arch is a toy.

The "Arch X" frame which is an instance frame of Arch has relatively few slots because most attribute values are inherited from class frame "Arch". The ISA slot indicates that the upper concept is "Arch", in other words, this is an instance frame of "Arch". The list "(a b c)" which is the value of the HASPARTS slot is a list of instances of (A B C) which is defined in the class frame for the arch. FINDSIZE is an attached procedure written in the LISP language to compute the size of "Arch X". It should be stressed that if a procedure is common to all instances of "Arch" the attached procedure should be defined in the "Arch" frame instead of in the "Arch X" frame.

Since the structure of the object is defined in class frame "Arch" by means of a combination of

HASPARTS and RELATIONS slots an inference for manipulating the descriptions must be managed by an extended inheritance control function. That is, since the structure of "Arch X" is not defined explicitly, the system has to infer it by substituting "a, b and c" by "A, B and C" respectively. This kind of substitution is not supported by ordinary frame-based knowledge manipulation systems and requires extension. Features relating to this function are installed as extensions to the ZERO system which process slots such as HASPARTS and RELATIONS.

MODELING OF COMPLEX STRUCTURED SYSTEM

In general a complex system consists of a number of subsystems, which in turn consist of a number of sub-subsystems, and can be decomposed until every subsystem becomes a primitive component. We need a method of representing such a system by means of an object model technique so that the complex system can be organized and manipulated systematically. We show that the method described in the previous section meets the requirement. By using a simplified automobile system a concept of representing the object model in the extended frame formalism and an example of representations will be shown followed by discussions about suitability.

The key point in the frame-based representation should be appropriate modeling so that the concept of prototype-instance (i.e., abstract -specific) relationships can be applied through the whole representation. This modeling method meets the requirements about the usage of the inheritance control which is the fundamental concept of the frame-based representation model. In other words, if a different modeling approach is needed, another type of knowledge representation model should be chosen. A frame-based system has its own inference control mechanism which supports only the inheritance handling based on the ISA relationship as discussed before. This means that knowledge models which are not based on the ISA relationship have no benefit from using the frame-based model. It should be noted again that a complex structured system can only be represented by a combination of abstract-specific relations and whole-part relations. The idea used to model the arch must be extended to the modeling of complex systems. This is the issue of this paper.

Next we discuss how complex systems can be modeled. Fig.6 shows an example of a representation of

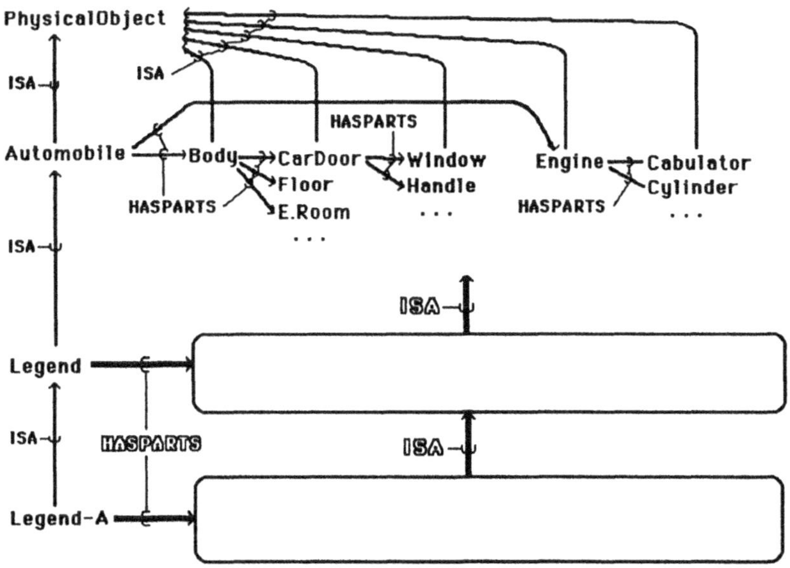

Fig.6 An example of a hierarchical representation of a complex structured object.

FRAME:	**Automobile**	
ISA	FRAME	PhysicalObject
HASPARTS	FLIST	Body, Engine, Steering, Electro
RELATIONS	PREDICATES	(INCLUDE Body Engine),
		(TOGETHER Engine Electro),

LENGTH	VALUE	NIL
---	---	---
---	---	---

FRAME:	**Legend**	
ISA	FRAME	Automobile
LENGTH	VALUE	465 cm
---	---	---
---	---	---

FRAME:	**Body**	
ISA	FRAME	PhysicalObject
PARTOF	FRAME	Automobile
HASPARTS	FLIST	Floor, E.Room, T.Room, ...
RELATIONS	PREDICATES	(FRONTOF E.Room Floor)

COLOR	TEXT	NIL
---	---	---
---	---	---

Fig.7 Frame descriptions for three components in Fig.5 based on the object model concept.

a simplified automobile system based on our modeling concept. For simplicity of discussion this model consists of four major subsystems, which are body subsystem, engine subsystem, steering subsystem and electronics subsystem. In addition, the body subsystem includes several sub-subsystems such as a floor, an engine room, a trunk room and a door which consists further of a window and a handle sub-subsystems. Similar decompositions are possible for other subsystems such as the engine. By means of this decomposition the automobile system can be decomposed until each module becomes a primitive part. The modeling should be done as follows. On the left hand side the principal objects starting with the physical object are defined by means of ISA relations, i.e., "Physicalobject -Automobile - Legend - Legend-A". This relation plays the role of the skeleton in the modeling, and may be called the skeleton hierarchy. On the right hand side the structures of the associated principal objects which are defined in the skeleton hierarchy are defined by means of HASPARTS relations. Each node in the hierarchy represents a frame as stated in the frame structure for the arch. The HASPARTS relations (and semantic network relations) are embedded within the parent system frame as discussed in the following.

It is very important to note that although the structures are defined around the principal objects at each level in the ISA hierarchy a set of of the structural definitions can be inherited from upper to lower levels, just as in the model description for the arch. That is, the structural descriptions of Automobile are inherited to those of Legend, which are inherited to its instance Legend-A. In this case, there is no need to redefine the structures at the second and third levels. If a different subsystem is used in Legend then definitions for that is needed. In this case the information to be inherited about the structural descriptions from Automobile to Legend-A is partly replaced. It should be noted that the components are represented by means of ISA hierarchies, and the structures are represented by means of HASPARTS hierarchies, and the structural informations can be inherited naturally by the extended inheritance control.

Fig.7 shows an example of frame-based representations for three objects from Fig.6 in the ZERO formalism, which are Automobile, Legend and Body. The last one is an example of non-principal object and includes a PARTOF relation to express that Body is a part of Automobile. It seems to be possible to represent a complex structured system as an object model according to the general rules by means of a frame-based formalism in a systematic way.

From the above discussions we can obtain the modeling rules for the systematic knowledge representation of complex structured objects, which conform to the concept of the object model:

1) The principal objects of a target system must be represented in a skeleton hierarchy by means of ISA slots.
2) The components are organized in an abstract-specific hierarchy using ISA slots, while the structures are organized in whole-part hierarchies using a combination of HASPARTS, PARTOF and RELATIONS slots.
3) HASPARTS slot has a list of subsystem element names as the slot value. The semantic relations between the elements are described by a set of predicates which is stored as the value of the RELATIONS slot of the frame.
4) Additional attributes of each component are described using user definable slots of an associated frame.

This modeling method is applicable under the extended frame-based knowledge representation environment. It should be noted that this knowledge representation formalism does not include heuristics, that the modeling does not need expertise, and that the knowledge seems to be application independent. In the following chapter we show several examples of the object-model based expert systems from different domains.

APPLICATION 1: TROUBLE SHOOTING SYSTEM COMBINING SURFACE AND DEEP KNOWLEDGE

Outline of System

As in Fig.8 this expert system consists of production system, i.e., surface knowledge system, and object-model system, i.e., deep knowledge system, e.g. Oomori and Ueno[14]. The production system is used to represent expertise in rules, and rule-based reasoning is applied to choose the most likely hypothesis or to decide on the unit to be focused to which the model-based reasoning will be applied. The reasoning manager controls the overall reasoning.

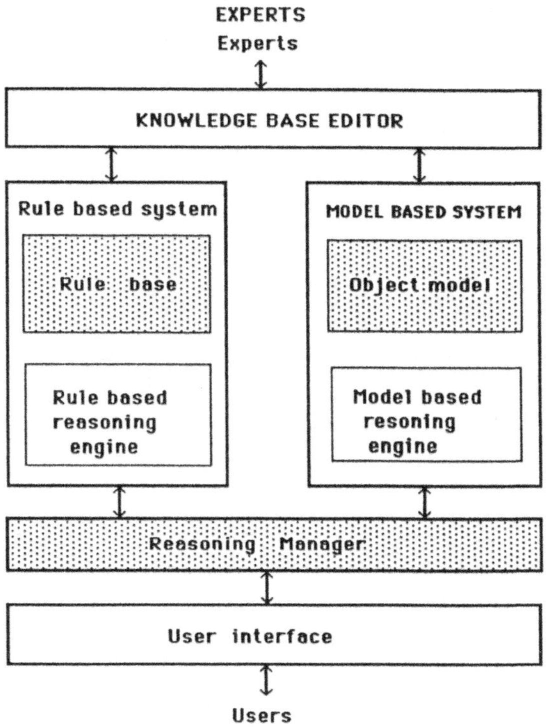

Fig.8 An overall view of a hybrid trouble shooting expert system combining a rule-based system and a model-based system.

Fig. 9 shows a function-based system structure of a focal plane shutter system, a sub-system of a reflex camera system, and its proper control signals. As in the figure the system receives two external input signals, one from a shutter dial unit and the other from a release switch unit, and gives an output to a mirror system. The signal from the release switch activates the action, that is, the control unit sends a signal to A-magnet which controls A-screen (front screen) by switching the signal from "on" to "off", and receives a signal from A-screen for confirmation. Similar signal is sent to B-magnet to control B-screen (rear screen). By difference of timing of the signals the shutter interval is appropriately controlled according to the input value from the shutter dial unit. The mirror is moved just before screen movement. Therefore, this is an example of a simple computer-based sequential control system.

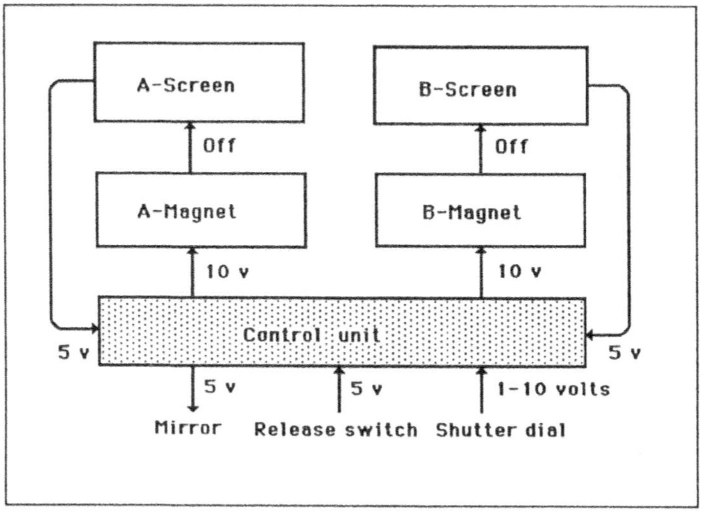

Fig.9 A function-based system structure of a focal plane shutter system and proper control signals.

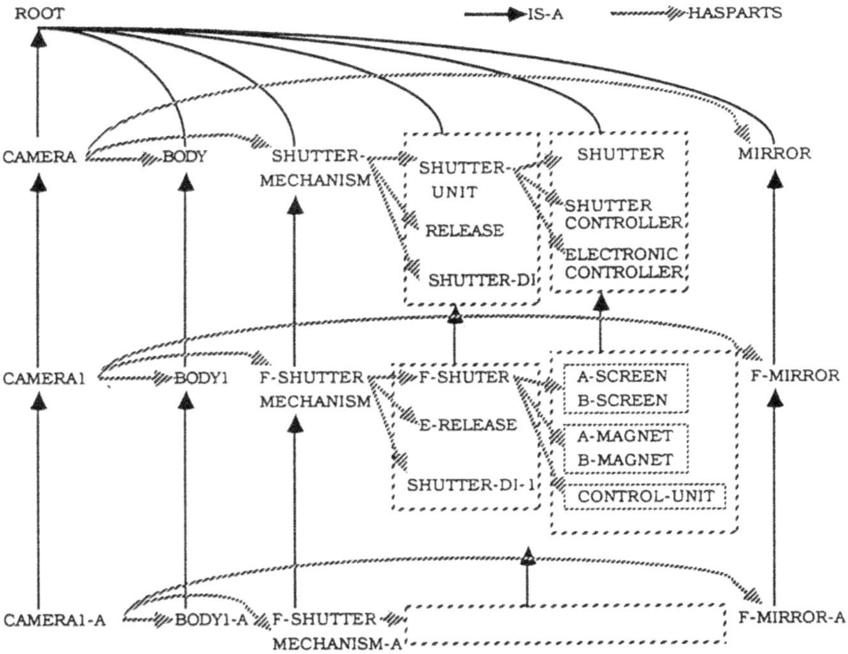

Fig.10 A hierarchical frame structure of a focal plane shutter system based on the object model concept.

Fig. 10 shows the hierarchical knowledge representation of the system based on the concept of the object model. As in the figure the hierarchy consists of three levels of representation. Camera, Cameral and Cameral-A are principal objects, and they constitute the skeleton hierarchy. Their structure is represented by HASPARTS relations at each level. At the top level the general structure of the camera system is described. This figure shows that Camera consists of Body, ShutterMechanism and Mirror, ShutterMechanism consists of ShutterUnit, ReleaseSwitch and ShutterDial, and ShutterUnit consists of Shutter, ShutterControl and ElectronicControl, respectively. At the middle level the specific focal plane camera named "Cameral" is described. As stated before if the structure and behavior of the focal plane shutter system is different from those of the general camera system the subdescriptions of the focal plane shutter mechanism named F.ShutterMechanism are redefined as in the figure. At the bottom level the instance object of Cameral, named Cameral-A, is defined, and the related frames are defined also. Since the values of the instance frames are same as those of the class frames between Cameral and Cameral-A all associated attribute data are inherited by means of the extended inheritance control features.

Fig.11 shows three examples of detailed frame descriptions from Fig.10. This figure shows two things; one is how the knowledge about structure and behavior are described within a frame, and the other is the relationships of the slot descriptions between frames in a hierarchy. The slots for structural description and their roles are:

A-kind-of	to describe the abstract object of this object in the ISA hierarchy,
HASPARTS	to describe a set of directly connected components of this object,
PARTOF	to describe the parent component of this object in the PARTOF hierarchy, and
RELATIONS	to describe the relations between the components which are defined in the HASPARTS slot of this frame.

The relations are described by means of a set of predicates. The slots for describing the knowledge about behaviors and their roles are:

BEHAVIOR	to describe the behavior of this object,

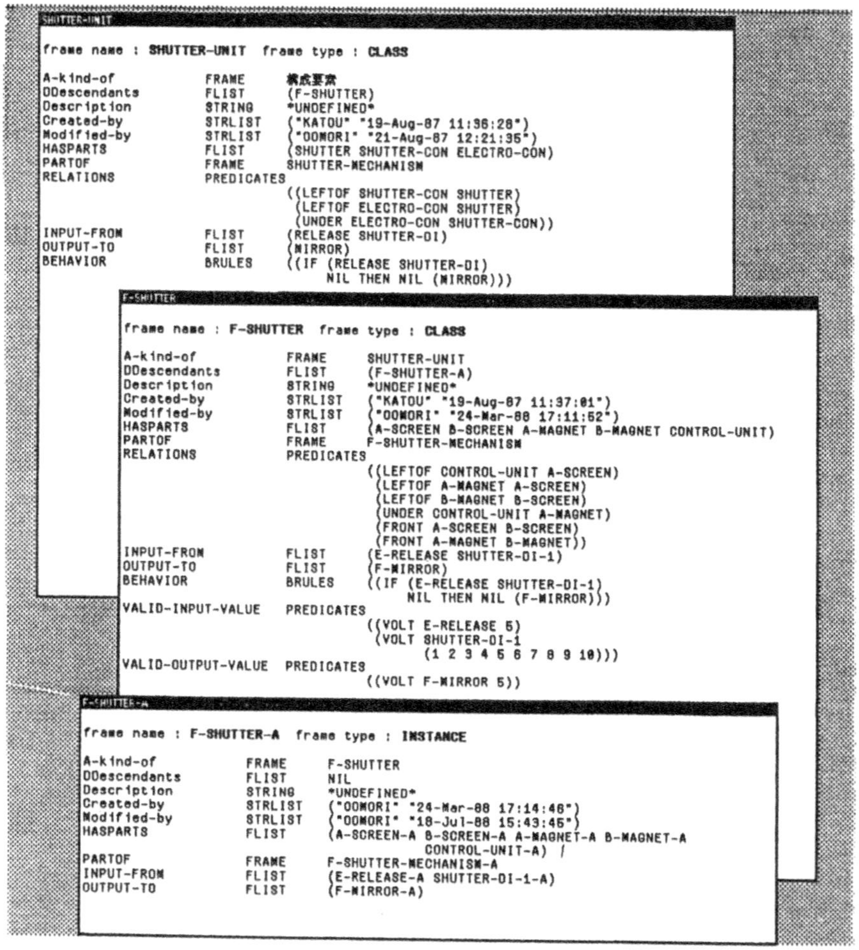

Fig.11 Detailed frame descriptions for three components in Fig.10.

INPUT-FROM to describe a set of elements from which the input is received,

OUTPUT-TO to describe a set of elements to which the output is sent,

VALID-INPUT-VALUE
to describe a list of valid input value ranges for each input, and

VALID-OUTPUT-VALUE
to describe a list of valid output value ranges for each output.

The behavior of an object, i.e., unit, is described by means of behavior rules in an IF-THEN rule form as in the following.

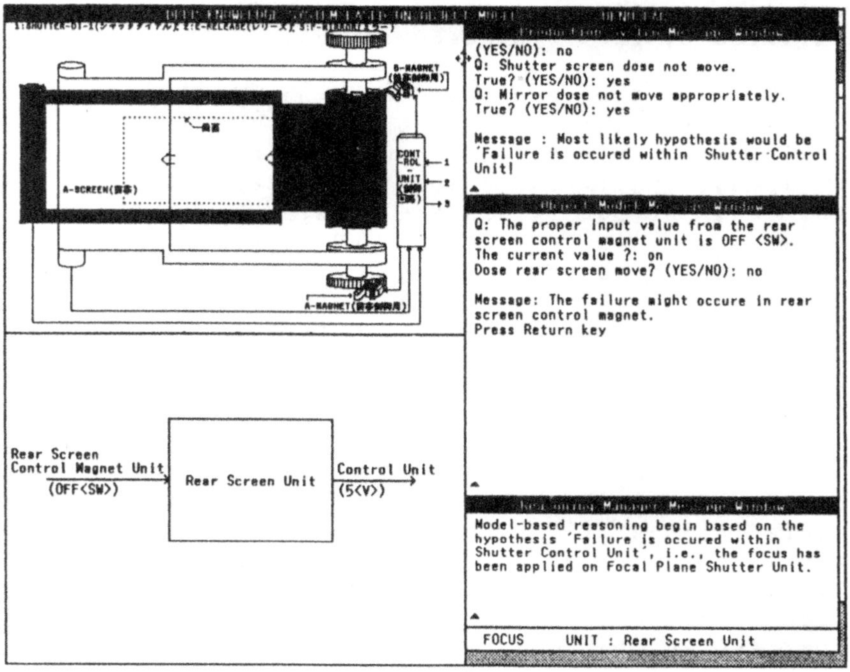

Fig.12 An example of displays of model-based interactive trouble shooting by the system.

IF input-condition, operations-to-this-unit
THEN actions-of-this-unit, output-to-other-units

The valid ranges for input and output are defined by means of a set of predicates as in the figure.

Although F-SHUTTER is the direct descendant, i.e., child, of SHUTTER-UNIT in an ISA hierarchy, the substructure of F-SHUTTER is different from that of SHUTTER-UNIT. Redefinition of structure and behavior about F-SHUTTER is done as in the middle frame. No such redefinition is seen in the bottom frame, since the data can be inherited from F-SHUTTER.

Inference is accomplished by combining rule-based reasoning and model-based reasoning. The rule-based reasoning which is done using a set of IF-THEN rules stored in the rule base of the production system is used to identify the focusing unit of the camera system. AND/OR tree-based backward reasoning is applied for this purpose. Subsequently the model-based reasoning starts from the focused unit which has been identified by the rule-based reasoning.

The strategy of the model-based reasoning is very simple. This is accomplished by repetitively checking the validity of input and output values of the focused unit. There are four cases; 1) both the input and output values are within the valid value ranges, 2) both the input and output values are outside the valid value ranges, 3) the input value is outside and the output value is inside, and 4) the input value is inside and the output is outside. In the first two cases the system concludes that the focused unit has no failures. The third case does not make sense. In the last case the system assumes that the focused unit includes the failure. If the focused unit has no subunits, or nothing is described about the subunits, then the system concludes that the failure occurs in this unit and terminates the reasoning. If there are descriptions about the subunits of the focused unit in the last case then the system continues the reasoning by changing the focus of attention to one of the subunits using the information from the associated behavior rule of the associated frame. The reasoning continues so that the most primitive unit is detected by this repetition. We think that this way of trouble shooting is quite similar to that of human experts. Fig.12 shows the interactive windows of the ELIS worksatation, displaying the contents of these frames.

APPLICATION 2: TROUBLE SHOOTING SYSTEM BASED ON OBJECT MODEL FOR PETROLEUM REFINERY PLANT

This example shows the object model of the refining system for propane and butane in an oil refinery plant, and a trouble shooting expert system by means of model-based reasoning, e.g. Kato, Oomori and Ueno[ ]. Although the detailed structures are different, since this is a chemical system with continuous and dynamic behavior, the basic modeling method and reasoning strategy are similar to those of the focal plane shutter system.

As shown in Fig.13 oil is pumped from the oil tank, heated by the heat exchanger and sent into the distillation tower. By difference in distillation temperature propane can be separated from butane within the tower. From the top of the tower gas, i.e., propane, is extracted and cooled by the condenser

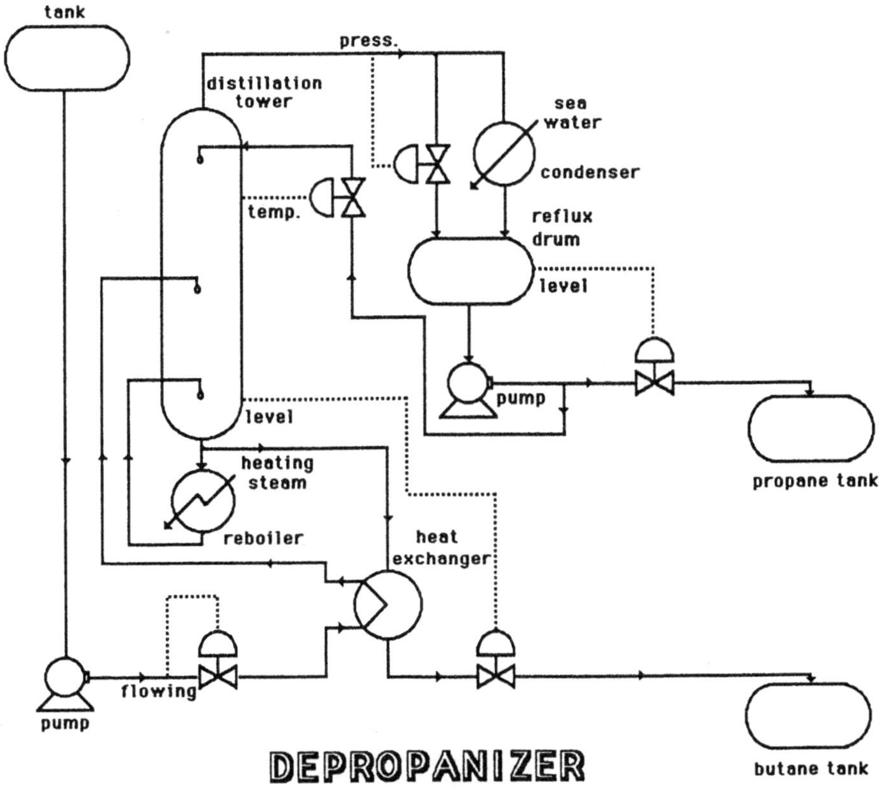

Fig.13 An overview of an oil refining system for propane and butane.

using sea water, and then sent into the reflux drum. Propane is pumped from the reflux drum and stored in the propane tank. Butane can be extracted from the bottom of the distillation tower, cooled by the heat exchanger which is used to heat the oil, and stored in the butane tank. The feedback system is installed for precise separation.

Fig.14 shows the hierarchical model structure. At the top layer the template structure of general refining subsystem is modeled. The refinery system is modeled at the second layer. The instance model is created and used for the model-based trouble shooting reasoning. In this case three attributes are examined which are temperature, pressure and quantity of flow. Fig.15 shows an example of interactive trouble shooting session of the system. The system is working on the ELIS workstation.

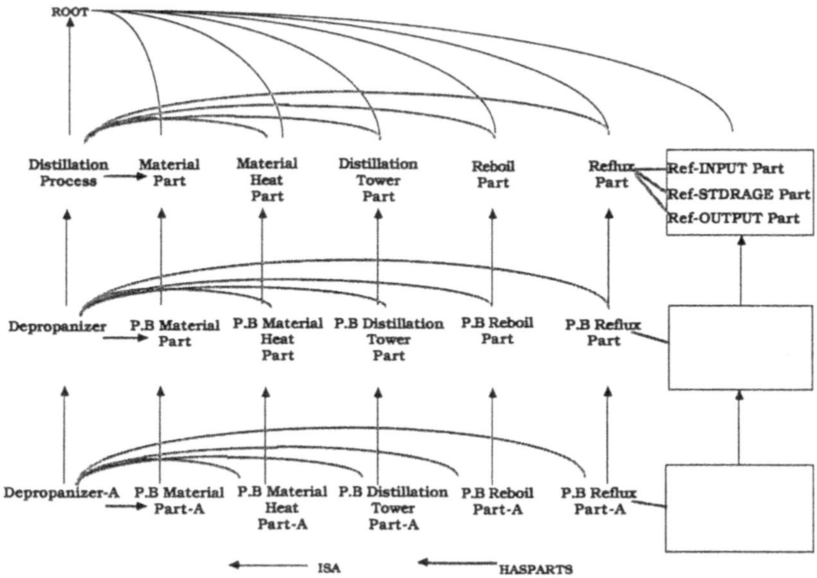

Fig.14 A hierarchical frame structure of an oil refining system based on the object model concept.

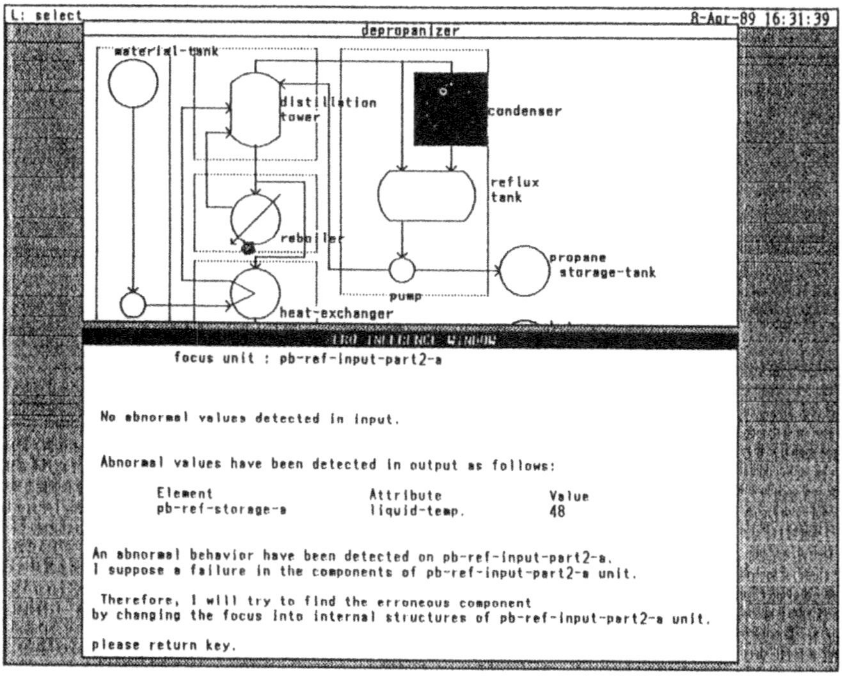

Fig.15 An example of displays of model-based interactive trouble shooting for an oil refining system.

APPLICATION 3: KNOWLEDGE BASED PROGRAM UNDERSTANDING

In this chapter the outline of knowledge-based program understanding system ALPUS is shown, e.g. Ueno[19,20]. ALPUS is a subsystem of INTELLITUTOR which is a knowledge-based intelligent programming environment for novice programmers. ALPUS comprehends user's programs using the knowledge about the algorithm associated with the program, detects logical errors and helps him fix them. The knowledge about an algorithm is represented by means of Hierarchical Procedure Graph which is a kind of object models, and the reasoning for comprehension is done by model-based reasoning. The detail structure of the model representation is, however, different from those which have been discussed so far.

Outline of System

Program understanding is an important theme both in AI and in knowledge-based software engineering, since "understanding" is in general a major issue in intelligent problem solving. However, few efforts have been tried on program understandings, in contrast to the amount of work done on automatic program generations and mathematical program verifications. It is easy to understand that program comprehension plays a key role in most phases of program development if one watches the actions of an expert programmer. Expert programmers use a variety of programming knowledge both in writing and reading programs.

A knowledge-based intelligent programming environment must be one of the most attractive and useful applications of AI concepts and technologies, e.g. Ueno[21]. By an intelligent programming environment we mean that a computer comprehends user's intentions from texts entered by him through an interactive terminal device, makes advices for fixing errors included, not only for syntactic errors but also for semantic ones, and helps him program more comfortably. These intelligent functions can only be achieved by a knowledge-based approach, i.e., an AI technology.

Advanced programmers can understand semantics of computer programs, detect logical errors, guess programmer's intentions from erroneous programs, and fix these errors. An intelligent programming assistant system is required to have these abilities. Knowledge about topics, algorithms, coding techniques and programming languages play a key role in this kind of situation. In this system the knowledge about algorithms plays a key role, and the knowledge about plans supports it.

Current version of ALPUS receives user's program written in the Pascal language. There are two types of understandings. One is "understanding the thing which one knows before". The other is "understanding the thing which one does not know before". Program understandings done by ALPUS is of the first type. That is, the system is able to understand the semantics of a program written by a user by means of the knowledge about the algorithm which is associated with the program. This is according to the assumption that a human tutor understands a program using his knowledge about the algorithm associated with it. Therefore, the knowledge about algorithms must be represented in a machine readable form.

From the point of view of the semantics in program design and program comprehension, an advanced programmer considers at least major variables and their roles, major operations on the variables and their roles, and a proper sequence of the operations. An identification of the specific role of each variable, of each operation on the variable, and the location of the operation in a program are, therefore, issues in program comprehension by a computer. An algorithm gives a major constraint to these elements, and for this reason we have chosen an algorithm-based program comprehension approach.

Knowledge Representations in HPG

Knowledge is needed to understand user's programs. ALPUS uses an algorithm-based programming knowledge. This knowledge is represented in a hierarchically structured graph called the HPG graph, Hierarchical Procedure Graph. The HPG graph works as a program template and consists of algorithm-related program attributes such as procedures, variables, statements and structural and semantic relationships among these components. A QUICKSORT program written in the Pascal language is used as an experiment. The QUICKSORT program was chosen because the algorithm was popular and useful in both programming education and business applications, and because the program was not too complex. There are two types of implementations, i.e., recursive procedures and non-recursive procedures. The non -recursive version is understandable for novice programmers. It is, however, not smart nor attractive in terms of programming techniques. The recursive version is relatively hard to understand for them, although it should be learned for the sake of advanced techniques.

The QUICKSORT algorithm is an algorithm to sort a random sequence of data which is stored in an array memory by a repetitive division of the sequence. By single division cycle one sequence of data is divided into a smaller value group and a larger value group. The division cycle is applied again and again so that every group member becomes single. The QUICKSORT algorithm gives a procedure, i.e., a sequence of operations on the array of data so that the initial state of the array is properly converted to the goal state. A set of variables to store the data, pointers and intermediate values and an operation sequence must be defined appropriately.

Fig.16 shows a skeleton of the HPG graph for SORT procedure of the QUICKSORT algorithm. As in the figure the SORT procedure consists of three major processes which are initialization process, search-and-exchange process and recursion control process, in this sequence. It should be stressed that the operations to achieve the QUICKSORT algorithm by the recursive processing technique are divided into three major groups and that operations related to these processes are not mixed together between different processes. This is an important difference between program comprehension and natural language comprehension although these two have several similar characteristics.

The initialization process is divided into three parallel subprocesses which are left-pointer-setting process, right -pointer-setting process and base-value-setting process. The search-and-exchange process, the second process, consists of two serial subprocesses which are search process and exchange-and-judge process in this order. The search process is a process to find two elements to be exchanged and is divided into two parallel subprocesses which are left search process for finding the first value element smaller than the base value from the left hand side of the array and right search process for finding the first larger value element from the right. The exchange-and-judge process is a process to exchange the two elements which have been found by the search process and then to narrow the width of the sequence by two and check the termination of the process. The subprocesses are structured as in the figure. The third major process which is the recursion control

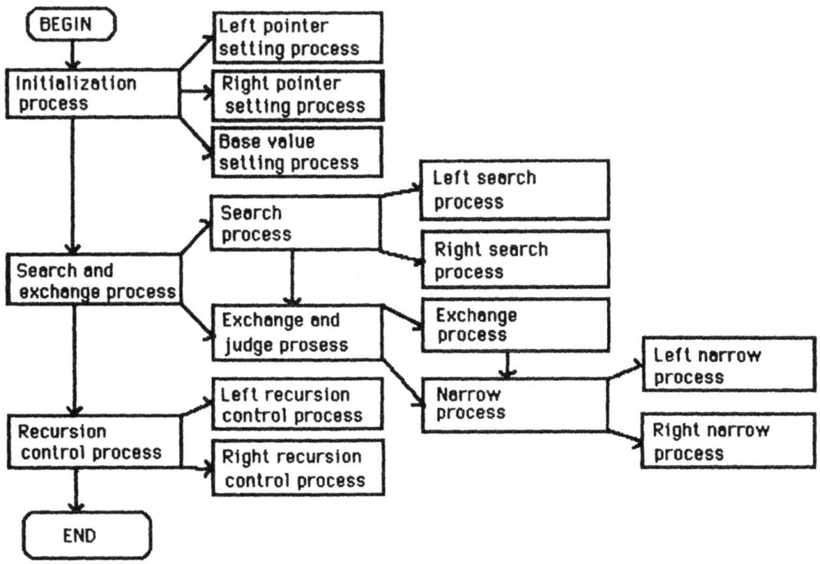

Fig.16 The HPG graph for the QUICKSORT algorithm.

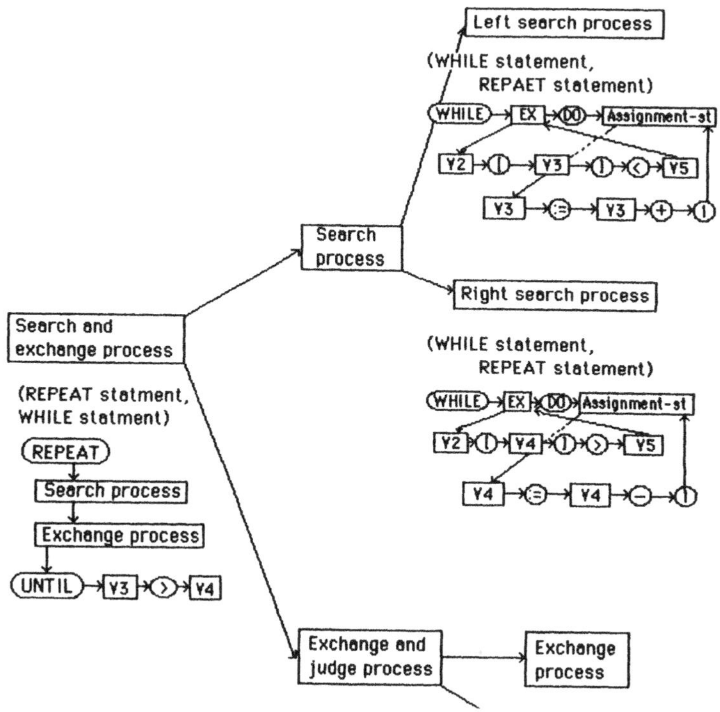

Fig.17 Attached information to the HPG graph.

process controls the recursions and consists of two subprocesses as in the figure.

The attribute informations for typical control strategies and operations are attached to each related process node of the HPG graph as in Fig.17. For example, for the search-and-exchange process REPEAT or WHILE commands are typically applicable. In the case of the REPEAT command a proper template is given as in the figure where v3(variable-3) and v4 are referenced as shown. These two variables appeared within the left search process and the right search process where other control strategies and operations are given as the templates as in the figure. These attribute informations are represented as abstracted procedure-oriented knowledge which is referenced by the knowledge-based pattern matching module of the ALPUS system in the program comprehension process.

In addition to the typical patterns of knowledge discussed above the HPG graph includes several acceptable patterns and error patterns. These additional informations could make the performance of the program comprehension more powerful and flexible. This knowledge have been extracted by cognitive psychology-based examinations applied to novice programmers at our school. Detailed discussions will be given in the following sections.

<u>Model Based Program Understanding</u>

The program understandings are done by knowledge-based pattern matching. That is, a user's program in an internal representation is checked against an associated program template represented in the HPG graph. Because variables that appear in the program would be different from those which are defined in the graph a reasoning to identify the role of each major variable is the first task. This task is done by combining bottom-up and top-down reasoning. Since ALPUS has the knowledge about the variables, which includes types of the variables, their roles and their locations to be referenced and to be handled, it is possible to identify the role of each variable by checking these attributes between the data, i.e., hypotheses, and the template, i.e., knowledge. Normalization of program representations is done prior to this task. Normalization is very important, but for the current version of the system the power of normalization is not sufficient, and the identification of an algorithm is rather simple. We are currently trying to develop a system that gives better performance. The following pattern matching processing is done by top-down reasoning. Since the knowledge base of the system that is the HPG graph includes several patterns for each technique which is attached to each process node in the graph, this reasoning does not require complicated processing techniques. Pattern matching is done by way of segment to segment matching. If the user's program segment matches the standard pattern, then the system understands this segment as a complete one. If it matches one of the acceptable patterns, then the system is able to make an advice to the user for writing better programs. It is also possible to rewrite the segment into the best one, a standard-like form. If it matches one of the error patterns then the system is able to assume the user's intention using informations attached to the pattern, and can make an advice for fixing the error or for letting him know his confusions, mistakes or misunderstandings. Further advices are given by the TUTOR system according to the output informations of this system. A part of comprehension will be discussed in more detail next.

If neither the standard pattern nor any available patterns match the user's code then the system assumes that logical errors exist in the user's program. In cases of simple errors such as missing of binary operators or condition codes the system can detect them using the standard or acceptable template patterns. In other cases the system tries to identify an error by matching the code to the error patterns registered in the template frame. If one of the error patterns matches then the system comprehends it as the logical error and assumes the user's intention associated with it. Messages are generated according to the matching test and displayed as an advice. If no error pattern matches then the system assumes that some errors exist within the program but their identifications failed. In this case error messages are not enough as advices.

Fig.18 shows an example of an error detection by the ALPUS system and advices produced by the system for a user. As shown in the figure the system shows the precise error location in the algorithmic structure of the program, the type of the error, the possible error cause, advices for fixing, and so on.

CONCLUSION

In this paper we have presented the concept of the object model, the modeling method based on the

```
ZERO : ACTIVATOR
1   PROGRAM QUICKSORT ( INPUT , OUTPUT )
2    VAR A : ARRAY [ 1 .. 100 ] OF INTEGER
3    I : INTEGER
4    PROCEDURE SORT ( LEFTPARAMETER , RIGHTPARAMETER : INTEGER )
5     VAR LEFTINDEX , RIGHTINDEX , BASE , WORK : INTEGER
6     BEGIN
7      LEFTINDEX := LEFTPARAMETER
8      RIGHTINDEX := RIGHTPARAMETER
9      BASE := A [ ( LEFTPARAMETER + RIGHTPARAMETER ) DIV 2 ]
10      REPEAT
11       WHILE A [ LEFTINDEX ] < BASE DO
12        LEFTINDEX := LEFTINDEX + 1
13       WHILE A [ RIGHTINDEX ] > BASE DO
14        RIGHTINDEX := RIGHTINDEX + 1
15       IF LEFTINDEX <= RIGHTINDEX THEN
16        BEGIN
17        WORK := A [ LEFTINDEX ]
18        A [ LEFTINDEX ] := A [ RIGHTINDEX ]
19        A [ RIGHTINDEX ] := WORK
20        LEFTINDEX := LEFTINDEX + 1
21        RIGHTINDEX := RIGHTINDEX - 1
22       END
23      UNTIL LEFTINDEX > RIGHTINDEX
24      IF LEFTPARAMETER < RIGHTINDEX THEN
25       SORT ( LEFTPARAMETER , RIGHTINDEX )
26      IF LEFTINDEX < RIGHTPARAMETER THEN
27       SORT ( LEFTINDEX , RIGHTPARAMETER )
28     END
29     BEGIN
30      FOR I := 1 TO 100 DO
31       READ ( A [ I ] )
32      SORT ( 1 , 100 )
33      FOR I := 1 TO 100 DO
34       WRITE ( A [ I ] )
35     END


"One logical error has been detected in this QUICKSORT program. "

The error occures within the [Right-search] process  in  the [Search]
 process  in  the [Search-and-exchenge] process  in SORT proceduer.

"Intentions infered from your statements, types of errors and suggestions for
 fixing the errors are as follows:"

It is infered that you might intend to search a bigger value than the BASE
 value from the right hand side for array A using the WHILE
 statement in line 13 . However, the search from the right hand side
 will be failed by this statement. The index variable RIGHTINDEX
 must be subtracted by one for this operation.
Therefore, it is recomended that you should exchenge the '+' operation to the
 '-' operation to fix the error.

" Causes of the errors are supposed as follows: "
1: You might copy line 12 for ease and change the variable names, and forget
the change of the operation.
2:Although you intended to write a correct statement you might make a careles
s mistake by considering only the form of a indexing operation and the variab
le names since the statement in line 12 is almost same.
```

Fig.18 An example of advices produced by ALPUS. (Messages are translated from Japanese sentences.)

concept. The modeling is suitably achieved by means of the extended frame-based knowledge representation formalism which has been implemented in ZERO.

The methodology shown here is useful in manipulating complex structured systems. The flexibility and power of the methodology is examined through three examples. The first application demonstrated the usefulness of combining surface and deep knowledge. The second application demonstrated the generality of the modeling. The third application demonstrated the multiplicity in usage of the object model. In addition to these examples we are developing a variety of applications, such as intelligent CAD systems, multiuse knowledge systems for microcomputer system development and maintenance support.

Since the modeling is systematic and needs no expertise, the concept and the methodology discussed here seem to be applicable to automatic knowledge acquisition from the documents produced by design systems.

ACKNOWLEDGMENTS

The works shown in this paper are supported in part by Science Research Foundation of Ministry of Education, Petroleum Engineering Center, NTT Software Institute, NTT Human Interface Institute and Tokyo Denki University Research Center. The author would like to thank members of Ueno laboratory of Tokyo Denki University for system implementations and discussions. The author would like to thank Kotaro Hosokawa for editing the draft.

REFERENCES

1. Ueno,H. Expert Systems, Information Processing (Jyohoshori), Vol.28,No.2,pp.147-157,1987(in J.)
2. Buchanan,B. and Shortliffe (eds) Rule-Based Expert Systems, Addison-Wesley, 1984
3. Minsky,M. A Framework for Representing Knowledge, in The Psychology of Computer Vision (P.H.Winston, ed), McGraw-Hill, 1965
4. Feigenbaum,E. The Art of Artificial Intelligence-Themes and Case Studies of Knowledge Engineering, Proc. NCC78, 1978
5. Ueno,H. An End-User Oriented Language to Develop Knowledge-Based Expert Systems, Proc. COMPCON FALL 83, pp.523-529, 1983
6. Ueno,H. An Introduction to Knowledge Engineering, Ohmsha, 1985(in J.)

7. Ueno,H Object Model for a Deep Knowledge System, Proc. 7th ICMCDM, pp.792-800, 1986
8. Chandrasekaran,B. and Mitta,S. Deep versus Compiled Knowledge Approaches to Diagnostic Problem Solving, Developments in Expert Systems(M.J.Comb ed.),pp.23-34, Academic Press,1984
9. Davis,R. Expert Systems: Where Are We and Where Do We Go From Here, The AI Magazine,3(2),1982
10. Hart,P.E. Direction for AI in the Eighties, SIGART, 79, 1982
11. Ito,H. and Ueno,H. ZERO: Frame + Prolog, LNCS, 221,pp.78-89, Springer-Verlag,1986
12. Ueno,H. and Ito,H. FMS: A Knowledge Representation Language Based on The Frame Theory, IECE AL82-64,1982(in J.)
13. Ueno,H. Concept of an Object Model, Digest IPSJ, Fall, pp1105-1106,1983(in J.)
14. Oomori,Y. and Ueno,H. Trouble Shooting System Combining Deep Knowledge and Surface Knowledge - Application of Object Model -, IECE AI87-30,pp.17-24, 1987(in J.)
15. Kato,H. Oomori,Y. and Ueno,H. Trouble Shooting Expert System for Petroleum Refinement Plant Based on Object Model, JSAI SIG-KBS-8803-1, 1988(in J.)
16. Ueno,H. Knowledge Based Intelligent Programming Environments -From The Point of View of Program Understanding-, Information Processing(Jyohoshori), Vol.28, No.10, pp.1280-1296, 1987(in J.)
17. Jonson,W.E. Intention-Based Diagnosis of Novice Programming Errors, Morgan-Kaufman Publishers,1986
18. Soloway,E. and Ehrich,K. Empirical Studies of Programming Knowledge, IEEE Trans. on Soft. Eng., Vol.SE-10, No.5, pp.595-609, 1984
19. Ueno,H. INTELLITUTOR: A Knowledge Based Intelligent Programming Environment for Novice Programmers, Proc.COMPCNO89 Spring,pp.390-395,1989
20. Ueno,H. A Knowledge Based Program Understanding for Intelligent Programming Environment for Novice Programmers, Proc. bit'89(Berlin),1989(to appear)

Trends in Expert Systems in Japan

Y. Koseki(*) and S. Goto(**)
(*)C&C Systems Research Laboratories
(**)C&C Information Technology Research Laboratories
NEC Corporation, 4-1-1 Miyazaki, Miyamae-ku, Kawasaki 213, Japan

ABSTRACT

Expert systems are considered to be the most successful application of AI (Artificial Intelligence). In this paper, trends in expert system developments in Japan are discussed according to recent surveys on expert system applications. Trends are presented in terms of their number, applications, sizes, hardwares, and development tools. Also, some selected expert systems are briefly introduced as case studies. Those systems are in the areas of diagnosis and maintenance, manufacturing process planning, VLSI design, computer animation and music arrangement.

INTRODUCTION

Dramatic advancement in modern computer hardwares has enabled us to apply AI (Artificial Intelligence) technology to real-world problems. Among diverse applications of AI technology, expert systems are considered to be the most successful ones. Since pioneering expert systems, such as MYCIN [5], were developed in universities in early 1970's, computer hardware cost-performance has increased significantly and has enabled us to use a personal computer with a few MIPS processing capability with a cost of a fraction of an automobile cost. Those cost-performance increase in computer hardware has accelerated developments of practical expert systems.

In Japanese AI market, almost all of the computer manufacturers have marketed expert-system developing tools in addition to their supports on the AI languages, LISP and Prolog. Those expert systems developing tools have been in the market for a few years and are available on various hardware and operating system environments, including main-frames, work-stations and personal computers. Using those AI languages and tools, most of Japanese enterprises have developed applications in diverse areas ranging from hardware diagnoses to financial consultations.

The latest surveys on expert systems in Japan [1, 2, 3, 4] show a significant increase in number of developed expert systems in Japan in the last few years. In the 1989

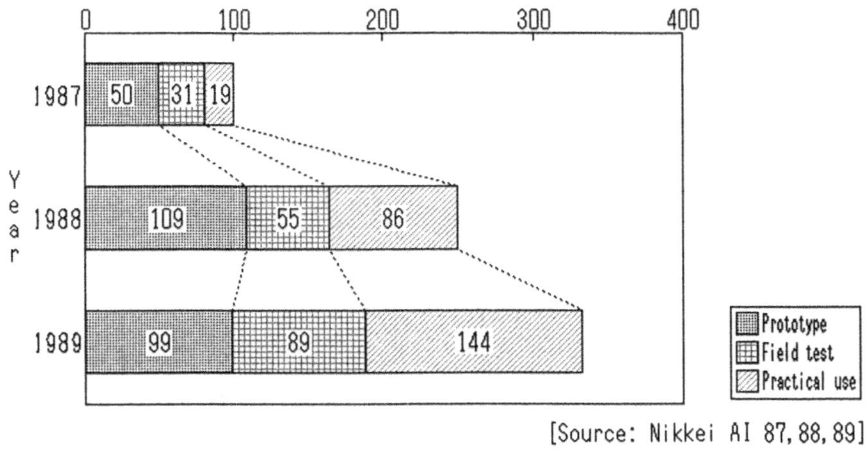

Figure 1: Number of Expert Systems Surveyed

survey, 332 non-AI-vendor developed expert systems are listed. Among those systems, 144 systems are claimed to be practically used. The ratio of practical systems against developed systems have been increasing year by year.

This paper analyzes the surveyed data and discusses trends in expert system developments in Japan. Trends are described in terms of their numbers, applications, sizes, hardwares, and development tools. After describing general trends, some selected expert systems will be described as case studies. Those systems are in the areas of diagnosis and maintenance, manufacturing process planning, VLSI design, computer animation, and music arrangement.

NUMBER AND SIZES

This section summarizes the surveyed data on Japanese expert systems [1, 2, 3, 4] and discusses general trends in their numbers and sizes. Figure 1 shows the increase in the number of expert systems listed in the surveys in the last three years.

In the 1987 survey, there were only 100 systems listed as non-AI-vendor developed systems. The number has increased dramatically in three years and it counts 332 in 1989. It is interesting to know that the number of prototype systems has decreased from 1988 to 1989, whereas the number of practical systems have increased significantly. The number of field-tested systems have also increased. This is because many of prototyped systems have been put into practice as time goes by. We may conclude that the expert system technology in Japan has evolved from an experimental phase to a practical-trial and practical-use phase. Figure 2 shows the progress of the ratio of practical systems among developed systems. The ratio of practical systems has increased from 19.0% of 1987 to 43.5% of 1989.

Figure 3 shows a distribution of the system sizes according to the survey of 1989.

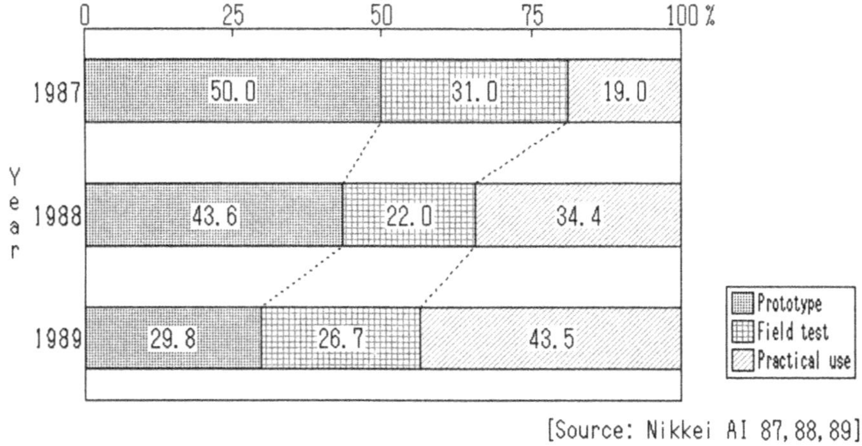

Figure 2: Ratio of Practical Systems

Sizes are classified in four levels: *tiny* (under 100 rules or equivalent), *small* (from 100 rules to 500 rules or equivalent), *medium* (from 500 rules to 1000 rules or equivalent) and *large* (over 1000 rules or equivalent). It is notable that almost half of the systems are tiny or small systems. If we suppose the same distribution for unknown cases, 71.6% of the developed systems are categorized as tiny or small systems. In practical systems, the ratio is slightly different. The ratio of medium and large ones in practical systems is greater than the one in overall. This shows that some of the tiny and small systems are developed for experimental or learning purposes, whereas medium and large systems are developed more seriously. It is also interesting to see that 49% of practical systems are tiny and small systems. As this analysis shows, it is still too early to consider the expert system technology in Japan is mature enough to be used in serious large-scale application systems.

DEVELOPING ENVIRONMENT

Figure 4 shows a trend in hardwares used in developing the surveyed expert systems. In this figure, *other machines* include Prolog machines [12]. The trend in hardware basically reflects a general trend in computer market. The use of personal computers and work-stations is steadily increasing. In contrast, the use of main-frames and mini computers is decreasing rapidly. Special hardwares for AI languages, that are Lisp machines, have become used less and less. It is naturally predicted that this trend will continue for a few years in the future.

It is interesting to see Figure 5 which shows the ratio of developing hardwares vs. delivery hardwares in the 1989 survey. Naturally, personal computers are the most preferable machines for delivery as long as the applications are small enough for them. Main-frames still seem to be dominant as delivery machines because they do not require further hardware investment. According to the survey, work-stations still seem to be considered as developing machines rather than delivery machines. However, it is predicted

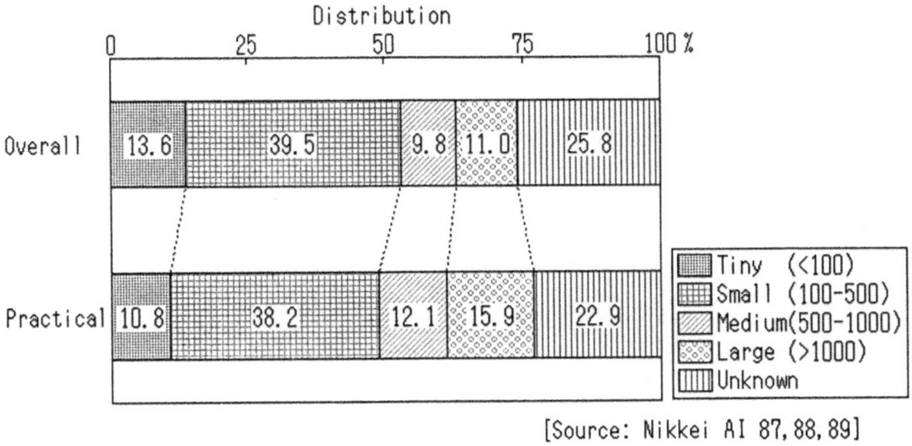

Figure 3: Sizes of Surveyed Systems (1989)

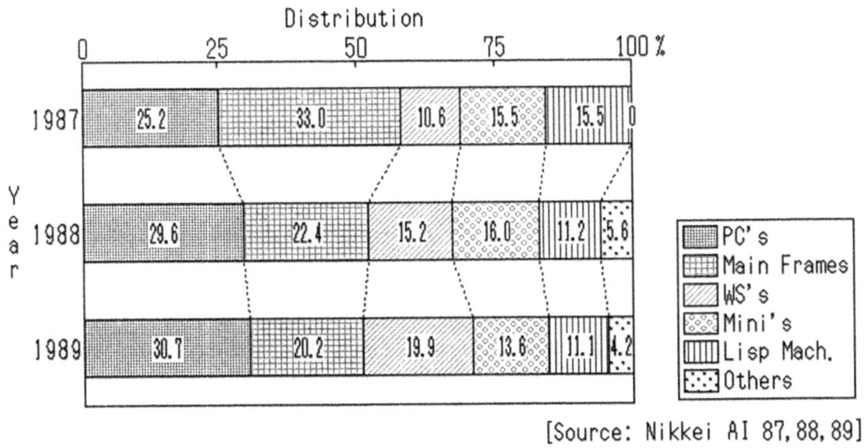

Figure 4: Developing Hardwares

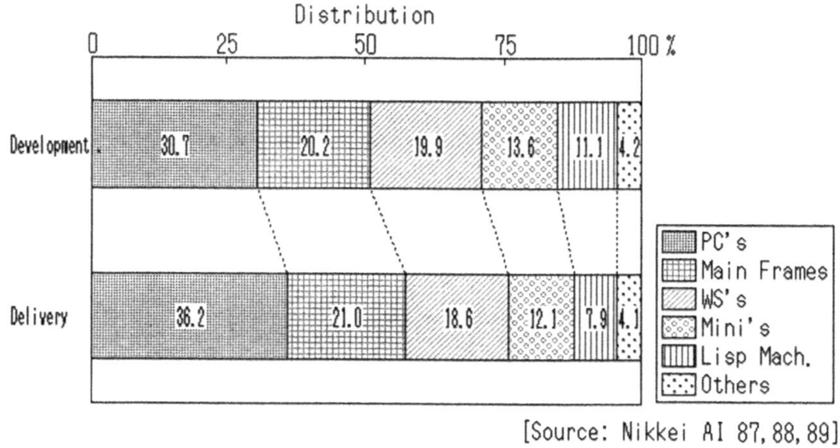

Figure 5: Hardwares (Development vs. Delivery)

that work-stations will be used more as delivery machines in the future as work-station prices decrease and work-stations become popular as general-purpose machines. Mini-computers and Lisp machines are less used as delivery machines and will be less used in the future.

Figure 6 shows a trend on developing tools. *Other tools* include specially developed tools for the application systems. Almost 60% of the systems are developed using commercially available expert-system developing tools.

APPLICATIONS

Figure 7 shows a distribution of the surveyed systems classified by types of developed companies. In this figure, the number of systems for electronics manufacturers is not large. This is because most of large electronics farms in Japan are also AI-tool vendors and their systems are excluded in those surveys.

In financial companies, 3 security companies, 10 banks and 2 credit companies are surveyed. They developed 28 systems altogether and put 16 of them into practice. Security company's applications include portfolio management systems (3), an online system terminal management system, a failure management of EDP system, an intelligent CAI system for option-traders and a support system for dealers. Most of banking farm applications are various consultations (loans, financial planning) for customers. They also have application systems in shop layout planning, automatic telex processing, budget planning and network diagnosis. A credit card company developed a system for credit-card application screening.

Systems developed for 15 insurance companies include insurance application screening, loan judgement, property management, medical insurance payment judgement, financial planning, risk evaluation, and various kinds of consultations.

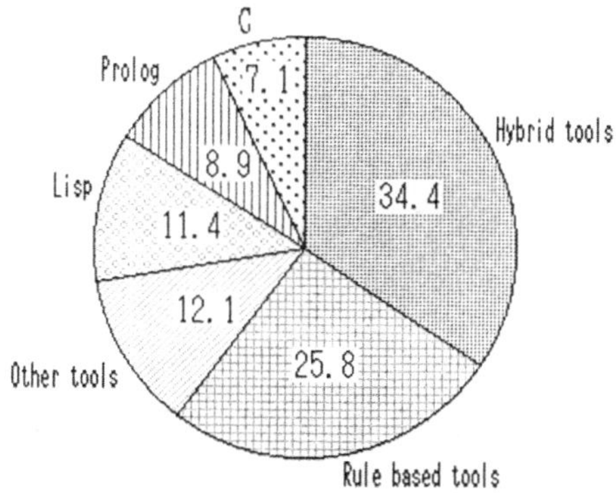

[Source: Nikkei AI 87,88,89]

Figure 6: Developing Tools

Marketing, service, distribution and transportation companies (14) have developed airline clue scheduling, airplane scheduling, instructor scheduling, resource diagnosis, communications network maintenance and management, and advertisement phrase design.

Steel and metal industries (8) have succeeded in putting most of developed systems into practical use. This is because they started building expert systems early and already out of experimental phase and in practical-use phase. Their applications are in the areas of diagnosis and maintenance, process planning and scheduling, resource management, and shipping scheduling.

Surveyed Construction companies (22) have developed 29 practical systems out of 73 developed systems. their applications range among building facility diagnosis, process selection, document production, process planning, and land-utilization consultation.

Electricity (4) and gas (2) companies have 24 applications including power plant diagnosis, power plant operation CAI, power plant operation and management, gas plant management, and network management and diagnosis.

Oil and chemical industries (20) have developed 33 systems for oil plant management and diagnosis, turbine control, chemical plant management, face make-up consultation, process management and diagnosis, and operation scheduling.

Shipbuilding farms (4) are using expert systems for chemical plant process consultation, plant online supervision, and chemical plant process design.

Mechanical and precision machinery manufacturers (23) are developing expert systems for mold design, lens design, LSI design, plant operation, and plant diagnosis.

Automobile industries (6) have applications in process scheduling, mechanical design, diagnosis consultation, and process planning.

Electronics manufacturers (5) have developed systems in diagnosis, software design, and hardware design.

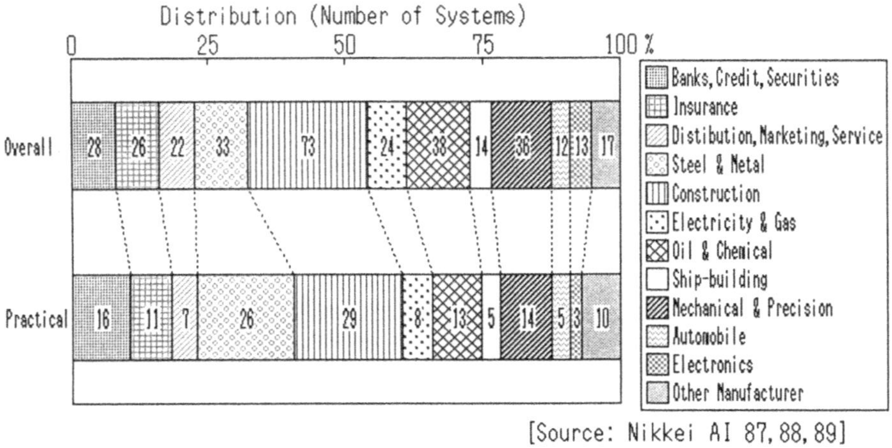

Figure 7: Classification by Type of Companies

Some of other applications are interesting. They are a wine selection consultation, a show-case layout, and a paper package design.

Figure 8 shows a trend in types of applications surveyed. It is obvious that the number of diagnosis and classification applications have been decreasing constantly. On the other hand, the number of planning and scheduling systems have increased dramatically. Because most of pioneering expert systems like MYCIN [5] in 1970's were diagnosis systems, early rule-based tools which were evolved from those systems were geared to diagnosis and classification tasks. In 1987, end-user companies were still rather in learning and trial phase in expert system technology. Therefore, it was natural to build expert systems for diagnosis and classification tasks. In fact in 1987, almost half of the developed systems were in those domains. After those companies learned much about the technology, they started developing practical systems with reasonable cost performances. It is conjectured that they have gradually found that it is not an easy task to build practical systems in diagnosis domain. The reason is that diagnosis systems themselves do not contribute to company profit directly. Those systems may contribute in reducing operation and maintenance expenses, but those contributions are hard to be proved. Another reason may be that the knowledge acquisition is very expensive task with today's shallow knowledge based expert system building style.

On the other hand, planning and scheduling domains look promising in building economically practical systems. This is because those systems directly contribute to company profit. It is too early to conclude that planning and scheduling domains are the most suitable for today's expert system technology, but it is predicted that applications in those domains will increase in a next few years.

Designing systems also seem to be promising, because it also directly contributes to company profits if they are capable of producing human-comparable designs. However, it seems that today's shallow knowledge based technology is not enough to build such systems. In fact, the ratio of practical designing expert systems has been decreasing in the last three years.

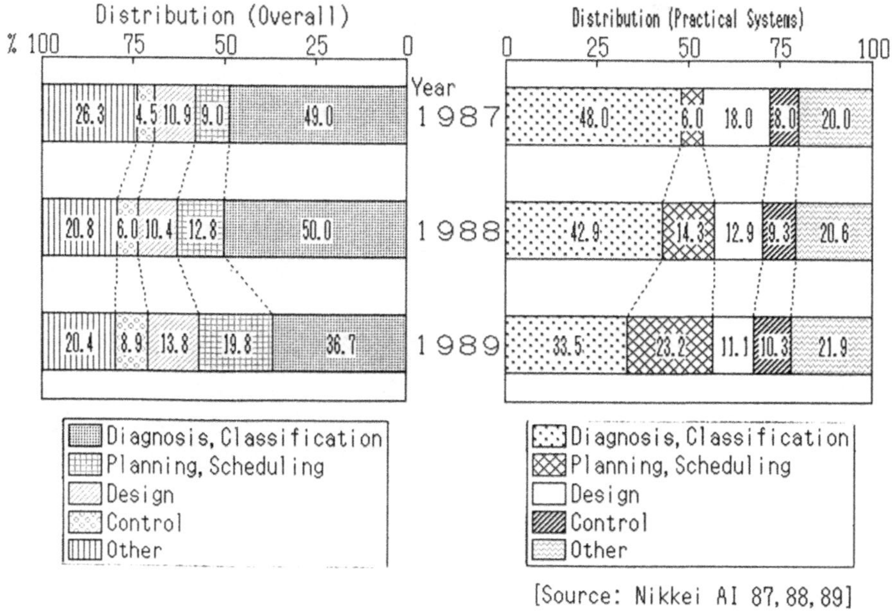

Figure 8: Type of Applications

CASE STUDIES

This section introduces some of experts systems developed in NEC Corporation as case studies of systems in Japan. Systems introduced include diagnosis and maintenance systems, a manufacturing process planning system, VLSI design systems, an automatic animation system, and a music arrangement system.

Switching System Diagnosis

Modern electronic devices like an ESS (Electronic Switching System) are very complex and difficult to diagnose. Although diagnostic test functions are built into ESS, defective components are not always clearly found by using them. By contrast, highly skilled human experts can locate defective components and repair them by using various diagnostic strategies. This has motivated development of a troubleshooting expert system called SHOOTX which modeled human experts' strategies [6, 8]. Some of the features of the developed system are as follows:

1. The system utilizes several heuristic test methods of human experts in order to locate the defective components.
2. The system has a kind of structural knowledge of the target ESS called an *abstract-signal-flow* model. The development and modification of the knowledge base in this style is far easier than those in diagnosis systems using empirical symptom-cause associating rules like in MYCIN [5].

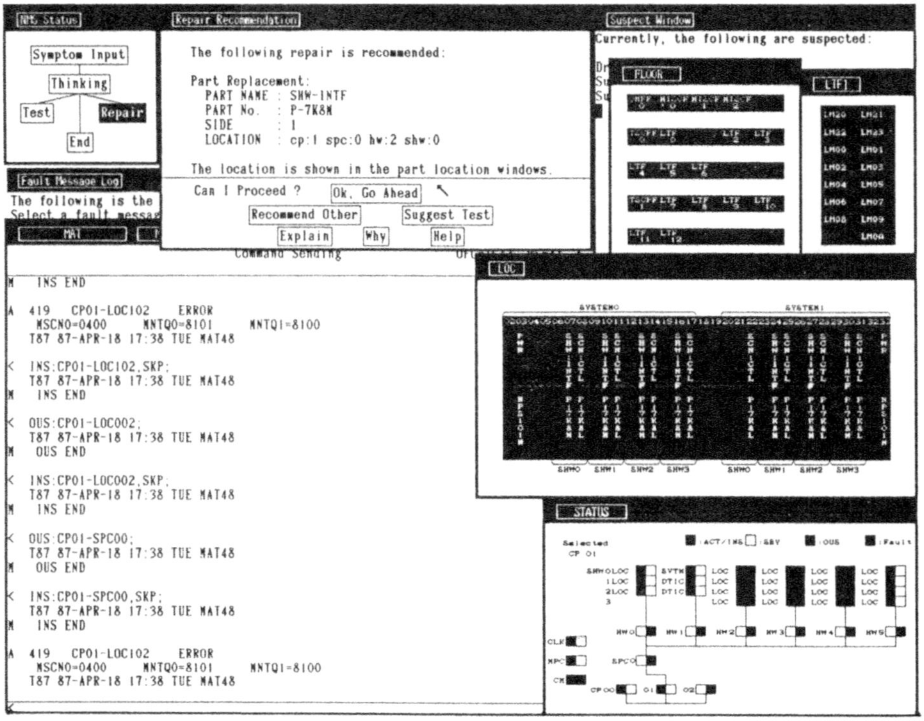

Figure 9: Switching System Diagnosis System SHOOTX

3. Several kinds of knowledge are represented in a unified network by utilizing object-oriented facilities. By this representation style, the knowledge base is kept compact and modifiable.

4. The system is designed to allow much flexibility in man-machine interaction. Therefore, an inexperienced technician can perform diagnosis just by following the recommended actions, while an experienced technician can freely choose alternative actions based on his judgment.

A prototype system with these features was developed on NEC EWS4800 series work-stations using a Prolog based tool called PEACE [7]. Figure 9 shows the user interface screen of the system.

Moreover, truth maintenance [11] facilities have been incorporated into the system in order to achieve indeterminate inferences. In the inferences of suspected components, there are many cases where definite conclusions cannot be made because of exceptions. The inferences done by human experts are heuristic and therefore might be erroneous. In order to narrow down suspected components, a kind of indeterminate inferences should be allowed. In order to make indeterminate inferences and properly revise them when they are known as erroneous, truth maintenance is necessary.

This new version uses assumptions to represent indeterminate factors included in diagnostic inferences, and it infers suspects based on them. Among many possibilities in selecting assumptions to use, the expert system adopts a control method which makes use of as many default assumptions as possible. By this control method, suspected components are narrowed down as much as possible, even if some contradictions are discovered. In addition, the method enables a reasoning based on anticipated test results without actually executing the tests. By truth maintenance and the control method, more flexible and effective diagnosis is achieved. A prototype system with this assumption based facilities was implemented on PSI Prolog machine [12].

Network Maintenance and Diagnosis

The prevailing trend in recent years has been toward the creation of communication networks with greater complexity, with a variety of host-computers, terminals and other equipments. This complexity has necessitated network management systems with increasing flexibility.

As the first step of a research on network management systems, a computer network fault analysis expert system called EXNETS (EXpert system for NEtwork Trouble-Shooting) [13] has been developed. EXNETS currently diagnoses faults in networks managed by a single host-computer. It will be enhanced to diagnose multiple host-computer networks in an integrated manner based on OSI network management protocols. The system is implemented using EXCORE [14] expert system developing tool on NEC EWS-4800 work-stations.

To determine types of faults to be dealt with the system, computer network experts have been interviewed by the questionnaire method. According to questionnaire results, the causes of computer network troubles are categorized into several types:

1. Hardware malfunctions.
2. Incorrect value settings of system generation parameters.
3. Incorrect user operations.
4. Software errors in user application programs.
5. Software errors in system programs.

Eighty percent of all troubles could be traced to types 1 to 3, which are targeted for diagnosis by EXNETS. The features of this system are as follows:

1. Applying to diverse network configurations (Hardware-independence)

   Network diversity, which involves a wide variety of network components and complexity of network configurations, has made it difficult to design trouble-shooting systems which apply to newly evolving networks. In designing the diagnostic strategy used in EXNETS, the reasoning mechanism is separated into two function modules. One module, called the Supervisor, generates hypotheses using common diagnostic procedures applied to faults of any network component. The other module, called the Executor, examines the network information collected from the individual network components according to the Supervisor's instructions.

   This fault analysis scheme in which the inference mechanism is separated into two function modules achieves several advantages in managing faults in diverse networks. For example, in a case where system configurations are altered, simple modification of the knowledge for the Executor is all that is necessary. Alteration of the whole system is not needed.

2. Providing intelligent assistance using knowledge of network experts

   Sophisticated diagnosis knowledge of network experts is an essential aid in managing complex network faults. The fault analysis configuration used in EXNETS has been designed based on trouble-shooting scenarios derived from the answers to the questionnaire, which detail personnel actions as they would likely be taken under given conditions. This information has been combined with data regarding the frequency as well as the potential significance of various problems to set up a set of certainty factors for checking out hypotheses in the fault analysis process. A fault tree which relates symptoms to causes has been defined in accordance with knowledge of network experts.

3. Implementing standard network management protocols (Compatibility with OSI management)

   In order to manage dispersed networks such as multivendor network environments in a unified manner, it is necessary to comply with standard protocols such as OSI management protocols. EXNETS uses C&C-DINA which is a family of NEC-designed protocols based on OSI standards. Management information between the Supervisor and the Executor is communicated based on Common Management Information Service and Protocol (CMIS/CMIP) in the OSI environment. The knowledge used by these diagnosis modules is stored in a Management Information Base (MIB). Using these protocols is believed to be effective in dealing with faults on multi-host-computer networks and multivendor networks.

Further Research on Diagnosis

There are two major disadvantages in conventional expert system approaches to diagnosis problems. The first problem is *brittleness* problem, that is, expert systems can only deal with prescribed cases in their knowledge bases. In other words, they are not robust for unexpected cases. The second problem is *knowledge acquisition* bottleneck, that is, acquiring experiential knowledge from human experts is difficult and time consuming task. To deal with those two kinds of major problems of expert systems, a novel method is being investigated by one of the authors [9]. The method combines so called *model-based* method (deep knowledge based) and expert system method (shallow knowledge based) to achieve robustness. It also has a capability to learn from its past experience.

Computer System Performance Analysis

Computer systems sometimes fail to attain maximum performance owing to insufficient hardware resources, inappropriately allocated resources, and unsuitable job scheduling strategies. To solve those performance problems, an expert system called EXPERFORM [15] has been developed. It can diagnose and propose counter-plans for performance problems on NEC ACOS-4 series main-frames.

The system exploits expert knowledge in a form of production rules, frames and predicates. Most of the effort on the study has been put on human expert analysis and organizing the knowledge structure. It is so well organized that a human expert, who has only one month experience in LISP programming, is able to write rules.

In production rules, diagnosis and counter-plan proposing knowledge is categorized into three layers, which are an investigation layer, a domain specific diagnosis layer and a hypothesis verification layer. By employing this structure, transferring expert knowledge to the system and maintaining knowledge base is kept simple and easy. To control rule-blocks, a blackboard mechanism is employed.

It has been released and has been evaluated in several user sites. As a result, its performance has been proved to be equivalent to that of an expert with three year experience. The system is implemented using EXCORE expert system developing tool [14], LISP and C.

Manufacturing Process Planning

Process planning is an activity which connects product design and manufacturing process. After the product design is finished, process planners transform given design information regarding mechanical components to manufacturing process information, such as machining tools to be used, operation sequences and cutting tool paths. ESPER (Expert System for Production EngineeRs) [16] was developed for automating process planning, that is, automatically generating manufacturing data, such as NC tapes, from CAD data.

Since process planning is essentially an optimization problem to minimize several pertinent criteria (e.g. total machining cost or machining time), problem solving techniques, developed in AI research, are effective in process planning to a certain extent. However, these techniques are insufficient for application to practical problems for the following reasons:

1. Since knowledge representation does not fully reflect the nature of geometry, planning is executed in a symbolized world, which differs to some degree from the physical world.
2. These approaches aim at establishing a single inference mechanism applicable to the overall process planning activity. As operation coverage widens, the system performance efficiency is reduced.

In order to compensate for the former issue about the physical world, geometric reasoning capability was introduced. For the latter issue about efficiency, a new optimization strategy and an interactive optimization method was introduced.

The geometric reasoning capability enables feature extraction from geometric data in the process planning viewpoint. Additionally, it allows direct manipulation of pertinent geometry while reasoning is going on. Representative manipulations of geometry are:

- Division and merging (to determine a shape of cutting area)

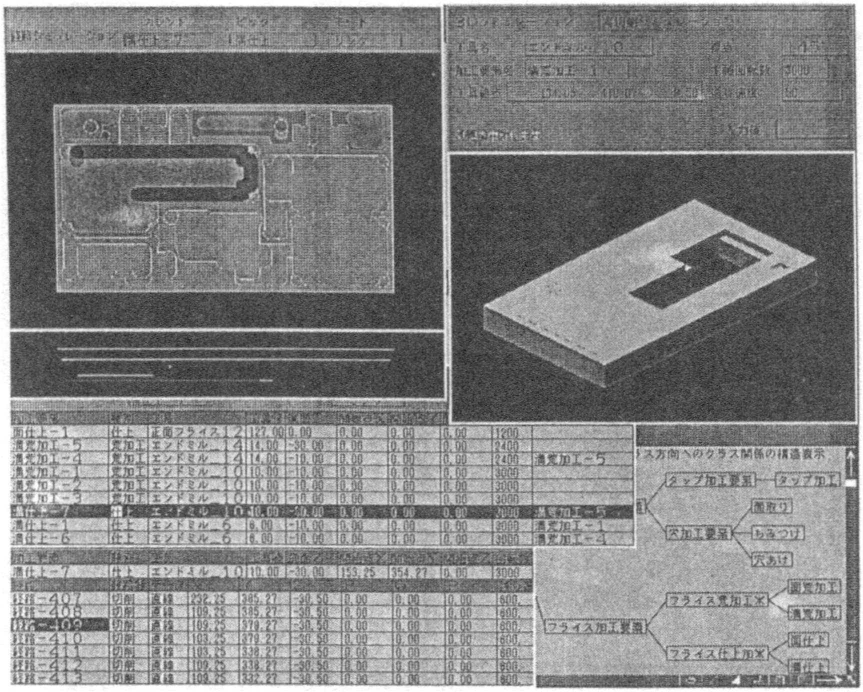

Figure 10: Process Planning System ESPER

- Subtraction (to combine rough cutting and finishing)
- Reduction (to generate a tool path)

Optimization requires effective problem solving strategy and, in addition, cooperation between the system and a user. For problem solving, knowledge is defined so as to avoid divergent search by pruning. Furthermore, user interactions can be incorporated into the optimization process to obtain a better solution.

ESPER automatically generates manufacturing data from CAD data. It also provides both graphical and tabular interface to a user, including cutting simulation, for comprehension ease. The system currently deals with 2.5 dimensional mechanical components which can be machined by a machining center, and is under evaluation for its effectiveness in practical use.

The system is implemented on NEC EWS-4800 using UTILISP and EXCORE [14], a knowledge representation language, featuring frame, ruie and object-oriented descriptions. The user interface is written in C. The system utilizes frame representation to describe design objects and manufacturing environments, while empirical knowledge is represented in the form of rule blocks. Figure 10 shows an user interface screen of the system.

VLSI Wiring Design

Routing is the most time consuming phase in the entire VLSI design process. From a computational point of view, a routing problem is considered to be a hard combinatorial problem. Further, it obviously is a large-scale problem, since the problem involves treating thousands of terminals of signal nets. An expert system called WIREX [17] has been developed for solving a routing problem in VLSI layout design and has been used to design actual VLSI's for several years.

This system has a hybrid configuration of combining an interactive CAD system and a knowledge base. The interactive system is written in Fortran and had been used extensively in designing gate arrays for several years. In addition to the interactive system, a special Prolog interpreter which has an interface to Fortran language was developed and a knowledge base was written as hierarchical rules with the language.

The interactive system consists of a heuristic algorithm-based automatic router which can complete most of the connections and an interactive graphical editor for human designer to modify the design and find routes for incomplete connections. After automatic routers have run, usually there are several incomplete connections left. Without the knowledge base, a designer had to examine a displayed layout on a graphic display and to come up with a strategy to modify the design to finish the incomplete connections. The knowledge base can take over this task. It applies rules written in its knowledge base to the situation and issues several editing commands to the interactive system. It asks some suggestions regarding local decisions from the user, and finally succeeds in connections. The system operates on NEC/MS mini-computers which was the machine for the interactive system. The system has been used in designing many gate arrays for several years.

Logic Synthesis for Gate Arrays

Because logic synthesis is one of the most important problems for quick and reliable VLSI design, many approaches have been proposed. They are divided into the following three groups, (1) Rule-based Approaches, (2) Transformation Table Approaches and (3) Algorithmic Approaches. Although the table transformation approach is suitable for technology specific mapping, the extent of optimization is limited. On the other hand, an algorithmic approach is a powerful tool for logic minimization problems; however, it is still weak for technology specific transformations. Therefore, it is desired that these two approaches be combined so as to utilize each advantage.

A logic synthesis system called FUSION [18] has been developed based on the combined approach, where not only tables for transformation are described as rules, but also a two level logic minimization algorithm is registered as one of the rules. A rule interpreter fires these rules using an effective branch and bound technique. Physical constraints such as longest path lengths between registers, fan-in/out and polarity are checked whenever each rule is applied.

The nature of the problem depends on the level of the input description. A functional description language called FDL [21] has been chosen as an input language which is a low level register transfer description language widely used in the company. Figure 11 shows an example description.

```
REG      FBSQ(0:1) = IF YD .UP. THEN
                     CASE FBBVTF-B8DVTF OF
                               /00/ NOC,
                               /01/ B8Q,
                               /10/ TQ,
                               /11/ TQ+B8Q;
```

Figure 11: Input FDL statement for FUSION

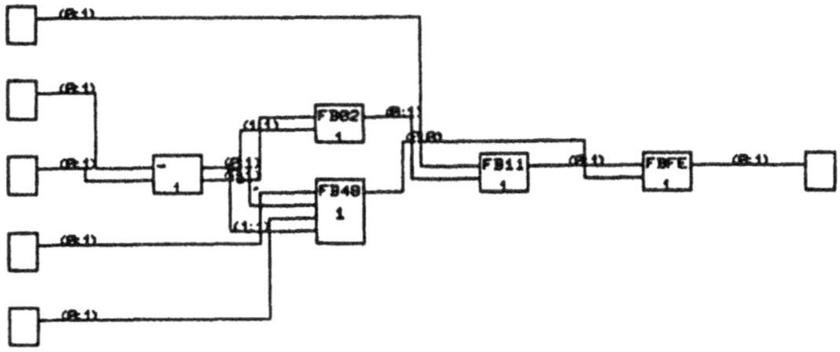

Figure 12: Result of FUSION

First, register-transfer descriptions are translated into a network. Technology independent optimizations such as constant propagation, removal of unreachable nodes, etc. are performed at this stage, using some algorithms. Then, the logical network is transformed into a physical one by firing rules and executing a logic minimization algorithm. The process is monitored on a CRT display. Finally, the final implementation data is generated. Figure 12 shows the final implementation data for the given example input. This output may be used as input data for layout and simulation. An interactive rule base editor is also available.

The system was implemented in the C language on EWS4800 work-stations and has been used to design actual gate array VLSI's with bipolar and CMOS technologies. Experimental results show that the combined approach is very effective and produces a solution very close to manual designs in a few minutes. They also show that the system with the logic minimization algorithm produces 20% smaller circuits on the average than the case where the algorithm was skipped.

Logic Synthesis for Full-Custom VLSI Design

The process of synthesizing a gate and switch level representation of a full-custom VLSI device from a given functional level description is still a largely manual process. It includes the most time consuming and error prone tasks in full-custom VLSI design. Full-custom design implies that, unlike gate array or standard cell design, little or no common structure is used to ease the design task. This motivated development of a knowledge-based CAD tool called EXLOG [19] for logic synthesis in full-custom VLSI design.

EXLOG accepts a specification of a functional module of a VLSI circuit, written in FDL (Functional Description Language) [21] and outputs a gate and switch level description in GDL (Gate Description Language) [22]. EXLOG produces the design through a process very similar to that used by design experts. As a result, it is able to use specific domain knowledge to construct and optimize circuits that would otherwise be intractable. It is a multi-paradigm knowledge-based system written in CL [20], a Lisp based expert systems building tool developed by the designers of the system. The primary characteristics of EXLOG are as follows:

(1) The EXLOG module refinement phase is closely modeled on the process used by expert designers. However, this process requires a higher level of abstraction than is provided by FDL. Therefore, EXLOG includes a module abstraction phase in which the given FDL is converted to a level of abstraction appropriate for the application of synthesis rules acquired from expert designers. Thus EXLOG is able to use expert knowledge in place of combinatoric search in the module refinement phase, and synthesize circuits similar to what a human designer would produce. This means that EXLOG needs two abilities. The first is the ability to partition the given functional level description. The second is the ability to implement specific pieces of logic. These abilities are achieved by utilizing expert knowledge of module abstraction and the integration of both basic circuits and custom-defined cells at arbitrary levels of abstraction.

(2) Additional information, which can not be written in FDL, is used to achieve a more sophisticated design. For example, EXLOG infers the role of signals (e.g. clock signals) and confirms them with the designer, asks the technology (e.g. CMOS, NMOS, or TTL) from which to fabricate the target chip, and asks the level of the speed. It also asks for timing information, such as when input signals are available, when output signals should be effective, and if there are any sets of input control signals which are exclusive at any time. This information is exploited in choosing optimal latch primitives, optimizing combinational logic circuits based on the required technology, and deciding the timings of precharging, charge-offing, and latching.

(3) EXLOG's implementation is based on the CL multi-paradigm programming system, where tasks can be partitioned into complementary subtasks requiring different programming paradigms. These subtasks are realized by using a frame/object based paradigm, a rule-based paradigm, a functional paradigm, or combinations of all three.

The EXLOG process consists of four main phases: FDL analysis, module abstraction, module refinement, and GDL generation. In the FDL analysis phase, given FDL statements for a functional module are translated into EXLOG's internal representations. At this phase, an environment module with external signals is generated and a macro cell library is loaded. These make up an initial knowledge base for the module being implemented. In the abstraction phase, sets of registers in a register transfer level description are grouped as special blocks where possible. In the refinement phase, module blocks at various levels of abstraction are refined by using both associated user-defined cells from a macro cell library and generic logical circuit elements (e.g. shifters, switches, and decoders). The generic module elements are then refined into boolean and switch level gates (e.g. NANDs, NORs, and pass transistors). This is done by selecting optimal latch primitives and optimizing combinational logic circuit networks. In the GDL generation phase, GDL statements are generated from the primitive module elements implemented in the previous phase.

The two-phase clock CMOS version of EXLOG was developed over a twenty month period. At present, the EXLOG rule knowledge bases total about 200 rules. This knowledge was acquired through interviews with design experts. Tens of example designs have been processed by EXLOG to produce complete CMOS full-custom designs. These designs have been verified by FDL/GDL simulations and by design experts. An EXLOG run takes about 5 CPU minutes on a SUN 3/260 for designs of functional modules of about 1000 gates. Compared to manually synthesized circuits, EXLOG's result was 83 % to 118 % in number of transistors. It takes about one and one half months for an expert designer to implement this size of design. EXLOG is starting to be used to design NEC's VLSI products including micro processors.

Circuit Selection

The circuit selection is one of the important tasks in VLSI designs. The circuit block selection problem is to derive a circuit which satisfies given requirements e.g. *synchronous binary counter which operates faster than 32 MHz and the power consumption of which should be less than 250 mw.* When there are no circuits which precisely meet the requirements, the most appropriate one must be derived by solving the trade-off situation between condition factors e.g. speed vs. power. Electronic circuit design experts mainly use "design rules" to select a circuit block which satisfies the given requirements. If there is no circuit block which precisely satisfies the requirements completely, the expert would interpret the design rules in a flexible way. Then he/she relaxes the pertinent conditions for applying the design rules and derives an acceptable solution.

In order to realize such flexible reasoning capability, a reasoning system with automatic rule relaxation has been developed [23, 24]. The system called EXPERT ESSENCE is an expert shell designed to meet practical design trade-off problems. Unlike conventional production system approaches, the system can automatically relax firing conditions of rules during the reasoning process. In the system, each rule has firing conditions. Here, a user can specify two kinds of conditions, i.e. necessary conditions and sufficient conditions. The necessary conditions must be satisfied to fire the rule. On the other hand, a sufficient condition can be relaxed. In the reasoning process, the system selects and fires a rule one after the other. If there are no rules which satisfy the firing condition, then the system tries to relaxes the sufficient conditions for each rule.

In the system, meta-rules are introduced to represent the design expert's knowledge on how to relax the firing conditions for each rule. Meta-rules represent the degrees to which each sufficient condition can be relaxed. They also represent differing levels of importance among condition factors e.g. speed and power.

In general, it is very difficult to construct a complete knowledge base. Even if a knowledge base for a design problem is constructed, it may give an unexpected conclusion for a certain requirements. In such a case, it is necessary to improve the knowledge base by a refinement process. To achieve this refinement process automatically, the system has an automatic knowledge acquisition mechanism. From a given set of expert designer's reasoning examples, the system generates appropriate meta-rules. It enables automatic knowledge-base refinements from examples at reasonable cost.

Rules are acquired directly from an expert and the meta-rule knowledge is acquired from given examples of an expert reasoning results. Therefore, in many cases, the user can refine the knowledge base without changing rules, by only adjusting meta-rules. For example, if the system recommended an unexpected circuit block for a certain requirement, an improved meta-rule can easily be obtained by using a new example set. Hence, the mechanism provides an ability to improve the knowledge base in a simple manner.

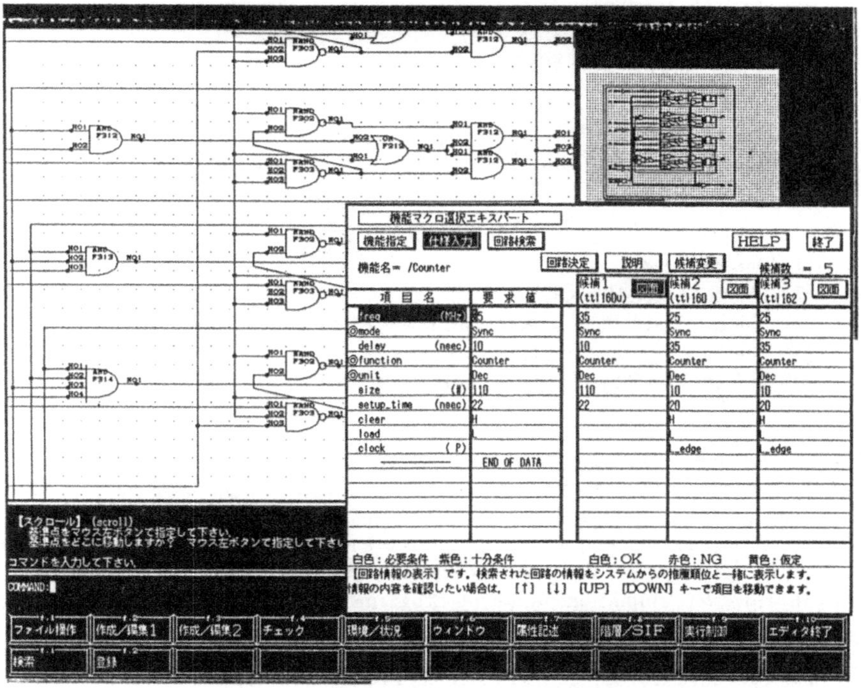

Figure 13: Trade-off Design System EXPERT ESSENCE

The system was implemented in C language on NEC EWS4800, M68030-based workstations. A knowledge base for 10 primitive functions are described in 800 rules. For 15 design examples of shift-register, the system could generate a meta-rule in about 2 minutes. The system has been in experimental use for designing VLSI circuits. Figure 13 shows an user interface screen of the system.

Story Driven Animation

Recently, computer animations have been widely used in many fields like conventional CAD, commercial films and movies. it is desirable to develop easy-to-use computer animation systems to encourage more wide-spread uses of computer animation. An animation generation system called SDA (Story Driven Animation) [25] has been developed to generate an animation for a given story written in Japanese taken from a children's story book automatically. The system reads a natural language story and creates its scenario for realistic graphic animations.

This system can accept curtailed expressions in input stories, so that story writers do not have to describe all the acts explicitly to get a desired graphic animation. There are three types of hidden actions which can be discovered by the system.

- Continuity of different actions between sentences

- Causality among actions
- Continuity of an action beyond a sentence

For example, we describe the first type of hidden actions which can be interpolated by the system. Consider the following sentences in "The hare and the tortoise" in Aesop's fables:

(1) The hare ran.
(2) The hare looked back.
(3) The hare said, "the tortoise can never catch up with me".
(4) The hare lay down on the grass.
("The Hare and The Tortoise", *Aesop's fable*)

Any person can conjure an accurate image of hare's actions. For instance, nobody thinks that the hare would lie down while running with his face looking in the reverse direction. If he were to do so, he would do a dive and his neck would be broken. Humans can naturally interpolate lacking facts between sentences with common-sense. In this case, we imagine that the hare stopped between (1) and (2), and looked forward between (3) and (4). However, animation producing programs without common-sense may produce strange graphic animations if these sentences are not explicitly described.

The SDA inference mechanism recognizes the need for interpolations of these hidden actions. A transition in a story is reflected by chronological sequences of multiple *worlds* each of which is monitored by TMS [26]. A world holds extracted assertions representing the situation inherent in a sentence. Each world is compared with its neighboring worlds, and assertions in the world are added/deleted/modified in the following three precesses:

- Causality check among actions.
- Interpolation of a continuous action beyond a sentence.
- Interpolation of hidden actions between neighboring sentences.

The system consists of three modules: 1) story understanding; 2) stage directing; 3) action generating. The first module reads a story written in Japanese and makes an action-based scenario. The second module receives the scenario and modifies it for stage setting. The last module generates animations on a graphics display according to the precisely specified scenario given by the stage directing module.

The knowledge base is constructed as action-oriented, hence it is independent of individual story domains. This is because knowledge related to actors' actions is common to everyone since actors' movements are constrained by physical limitations of human bodies.

The SDA story understanding system is implemented using Prolog on vax11/780. Currently, the system works well for several fables. Figure 14 shows a scene from an animation generated by the system for the story "The Hare and The Tortoise".

Figure 14: An Animation Generated by SDA

Music Arrangement

With the advent of personal computers and digital music equipments, such as digital music synthesizers, automatic music arrangements by computers are becoming possible. A music arrangement system, which can transcribe hummed melody into musical notation and can generate accompaniments automatically, has been developed. The system contains a NEC PC-9801 16-bit personal computer, a digital signal processor, an A/D (analog to digital) converter, a microphone and a music synthesizer.

It consists of a transcription module [27], a chord generation module, a melody analysis module and an arrangement module. The transcription module analyzes pitch and strength of hummed melody and transcribes it into music score. Two kinds of chord generation modules have been developed. One uses rule-based approach, while the other takes neural-network [28] approach. The neural-network-based one uses a three-layer feed-forward network which has been disciplined to harmonize given melody by back-propagation method. The melody analysis module examines an entire melody line and extracts melody structure. The result is taken by the arrangement module, which produces an accompaniment, an introduction part, an intermission part and an ending part.

CONCLUSION

This paper has described trends in expert system developments in Japan, according to recent surveys on expert system applications. Trends are presented in terms of their number, applications, sizes, hardwares, and development tools. Also, some selected expert systems are briefly introduced as case studies. Those systems are in the areas of diagnosis and maintenance, manufacturing process planning, VLSI design, computer animation and music arrangement.

ACKNOWLEDGMENTS

The authors would like to thank people who have willingly contributed the materials on their developed expert systems. The authors also would like to thank Yasuo Kato, Tatsuo Ishiguro, Yoshihiro Nagai and Hajimu Mori of NEC Corporation for various kinds of supports.

References

[1] "Special Report on Expert Systems," *NIKKEI ARTIFICIAL INTELLIGENCE*, Jan. 19th (Supplement), Nikkei-McGraw-Hill, Tokyo, 1987 (in Japanese).

[2] "Four kinds of characteristics on hundred expert systems in Japan" *NIKKEI ARTIFICIAL INTELLIGENCE*, pp. 16–17, Mar. 2nd, Nikkei-McGraw-Hill, Tokyo, 1987, (in Japanese).

[3] "Special Report on Expert Systems," *NIKKEI ARTIFICIAL INTELLIGENCE*, Jan. 18th (Supplement), Nikkei-McGraw-Hill, Tokyo, 1988 (in Japanese).

[4] "Special Report on Expert Systems," *NIKKEI ARTIFICIAL INTELLIGENCE*, Jan. 16th (Supplement), Nikkei Business Publications, Tokyo, 1989 (in Japanese).

[5] E. J. Shortliffe, *Computer Based Medical Consultations: MYCIN*, Elsevier, New York, 1976.

[6] Y. Koseki, S. Wada, T. Nishida, and H. Mori, "SHOOTX: A Multiple Knowledge Based Diagnosis Expert System for NEAX61 ESS," *Proc. International Switching Symposium 87*, pp. 78–82, Phoenix, 1987.

[7] Y. Koseki, "Amalgamating Multiple Programming Paradigms in Prolog," *Proc. IJCAI-87*, pp. 76–82, Milano, 1987.

[8] S. Wada, Y. Koseki, and T. Nishida, "An Assumption-based Fault Localization Technique for Switching System Diagnosis," *Proc. of The Second International Conference on Industrial & Engineering Applications of Artificial Intelligence & Expert Systems*, Tullahoma, June 1989.

[9] Y. Koseki, "Experience Learning in Model-based Diagnostic Systems," *Proc. IJCAI-89* (to appear), Detroit, Aug. 1989.

[10] T. Chikayama, S. Takagi, and K. Takei, "ESP - An Object Oriented Logic Programming Language," *ICOT Technical Report* TM-0075, Institute for New Generation Computer Technology, 1984.

[11] J. de Kleer, "An Assumption-based TMS," *Artificial Intelligence 28*, pp.127-162, 1986.

[12] K. Taki, M. Yokota, A. Yamamoto, H. Nishikawa, S. Uchida, H. Nakashima and A. Mitsuishi, "Hardware Design and Implementation of the Personal Sequential Inference Machine (PSI)," *Proc. of the International Conference on Fifth Generation Computer Systems (FGCS-84)*, pp.398-409, 1984.

[13] T. Yamahira, Y. Kiriha, and S. Sakata, "Unified Fault Management Scheme for Network Troubleshooting Expert System," *Proc. of International Symposium on Integrated Network Management* (to appear), May 1989.

[14] M. Yamada, Y. Aori, T. Ushijima M. Yamamoto, and M. Watanabe, "Expert Systems Development Tool EXCORE," *NEC Research & Development*, No. 87, pp. 76–85, 1987.

[15] M. Asogawa, A. Shiramizu, K. Ikumi, and H. Kubo, "A Performance Improvement Expert System for ACOS-4 Operating System," *NEC Research & Development*, No. 87, pp. 86–94, 1987.

[16] Y. Maeda and K. Shinohara, "Geometric Reasoning and Organized Optimization for Automated Process Planning," *Proc. AAAI-88*, pp. 105–110, St. Paul, 1988.

[17] H. Mori, K. Mitsumoto, T. Fujita, H. Wakata and S. Goto, "WIREX: VLSI Wiring Design Expert System," *Proc. of the IFIP TC 10/WG 10.5 International Conference on Very Large Scale Integration (VLSI 85)*, pp. 303–312, Tokyo, 1985.

[18] T. Yoshimura and S. Goto, "A Rule-based and Algorithmic Approach for Logic Synthesis," *Proc. of the IEEE International Conference on Computer-Aided Design*, pp. 162–165, Santa Clara, 1986.

[19] M. Watanabe, N. Maeda, M. Iwamoto M. Kawarabayashi, et al, "EXLOG: An Expert System for Logic Synthesis in Full-Custom VLSI Design," *Knowledge Based Expert Systems in Engineering: Planning and Design*, pp. 315–329, Computer Mechanics Publications, 1987.

[20] M. Watanabe, T. Yamanouchi, M. Iwamoto and Y. Ushioda, "CL: A Flexible and Efficient Tool Kit for Constructing Knowledge-based Expert Systems," *IEEE Expert* (to appear), August, 1989.

[21] S. Kato and T. Sasaki, "FDL: A Structural Behavior Description Language," *Proceedings of the IFIP 6th International Symposium on Computer Hardware Description Languages and Their Applications*, pp. 137–152, 1983.

[22] H. Sakuma, Y. Fujinami and T. Kurobe, "NELSIM: A Hierarchical VLSI Design Verification System," *Proceedings of the IFIP TC 10/WG 10.5 International Conference on Very Large Scale Integration*, pp. 109–118, 1983.

[23] T. Fujita and Y. Nakakuki and S. Goto, "A New Knowledge Based Approach to Circuit Design," *Proc. of the IEEE International Conference on Computer-Aided Design*, pp. 456–459, Santa Clara, 1987.

[24] Y. Nakakuki and T. Fujita, "EXPERT ESSENCE: A New Expert Shell with Automatic Rule Relaxation," *Proc. of International Workshop on Industrial Applications of Machine Intelligence and Vision (MIV-89)*, pp. 199–204, Tokyo, Apr. 1989.

[25] H. Shimazu, Y. Takashima and M. Tomono, "Understanding of Stories for Animation," *Proc. of International Conference on Computational Linguistics (COLING-88)*, pp. 620–625, Budapest, 1988.

[26] J. Doyle, "A Glimpse of Truth Maintenance", *Artificial Intelligence: An MIT Perspective*, The MIT Press.

[27] S. Tsuruta, M. Fujimoto, M. Mizuno and Y. Takashima, "Personal Computer Music System – Sung Song Transcription and Its Applications –," *Digest of Technical Papers, IEEE International Conference on Consumer Electronics (ICCE)*, pp. 114–115, June, 1988.

[28] D. E. Rumelhart and J. L. McClelland, *Parallel Distributed Processing* Vol. 1 and 2, MIT Press, 1986.

Computer Graphics and its expansion to image industry

M. Ishii and K. Murakami

Fujitsu Laboratories Ltd. Kawasaki 1015, Kamikodanaka, Nakahara-ku, Kawasaki, 211, Japan

ABSTRACT

Major parts of computer graphics techniques such as rendering and modeling, are presented. First, we discuss the current status of computer graphics and image industry in Japan, and point out requirements that solve problems which image industry is facing to. Next, we introduce several ray tracing methods we developed; *the fast ray tracing, the parallel ray tracing,* and *the incremental ray tracing.* With these method ray tracing may become practical for some computer graphics applications, such as CAD and animation which require high-quality images. For the modeling scheme, we developed a new framework, called *FUSION,* in order to integrate several types of processing for computer graphics into a multi-window environment. With this modeling scheme, large and complex objects are easily handled.

INTRODUCTION

Today, computer graphics are being used in a wide range of application including TV commercials, business graphics, computer-aided interior design, and visualization of computer tomography images or simulation. In Japan, practical-scale video production using computer graphics began at the Tsukuba EXPO held in 1985. Before that, some trials had been run, but they were far from practical. On display at the exposition were several three-dimensional shows, including one by Fujitsu Omnimax entitled "The Universe". Many other productions followed. Many computer graphics shows will be on display at the EXPO '90 to be held in Osaka. Fujitsu intends to display full-color three-dimensional images on a 24-meter omnimax screen. These images will show how plants store solar energy by photosynthesis and how animals use that energy by eating plants. Other than those applications, one of the major construction companies developed a ray tracing system for design and are applying it to exterior and interior simulation. In this way, the Japanese video industry is about to make new developments.

So far, the computer graphics industry is still very small, and is mainly geared to exhibitions. There are no graphics companies that have more than 50 employees. And most are very small. This is probably because the business is not very profitable, and there is not enough investment to develop new technology. To solve this problem and help develop the market, major Japanese computer makers such as Fujitsu and Toshiba are joining forces with various universities to research and develop computer graphics.

This paper explains the current status of computer graphics in Japan, focusing on the efforts made by Fujitsu.

The computer graphics industry is now facing technical problems such as low rendering speeds, increasing labor for modeling, difficulties in creating animation and providing a general-purpose database. Requirements to solve these problems are as follows:

Rendering

Both speed and quality are needed for rendering images. The Z-buffer algorithm provided speed. Since the Z-buffer algorithm is fixed, many graphics engine which use custom LSIs and DSPs are being developed. To make high-quality images Hiroshima University is researching a rendering method based on radiosity. Ray tracing is also used to generate high-quality images, but it takes a long time. Several weeks' work go into a few minutes of animation. The hardware and algorithm does not meet the requirement and is still at the research stage. NTT is developing high-speed algorithm, called pencil ray tracing. Fujitsu has also conducted research on high-speed ray tracing and has successfully developed the method explained in section 2.

Reduced labor for modeling

Computer graphics also require a modeling system with a high-level user interface. Modeling techniques mostly originated in CAD technology. A professional operator is assigned for CAD shape input, but computer graphics are usually processed by an artist not accustomed to using a computer. Moreover, shapes used in computer graphics is different from one in CAD, and input method is also different. To solve this problem, a new modeling method must be developed.

In the Japanese automotive industry, some makers have developed their own CAD systems, and connected them to supercomputer for structural analysis. Some electrical manufactures are also designing products by using their design systems. However, there are no general-purpose computer graphics modeling systems.

Animation

Most of the computer graphics shows involve animation. Although the key frame method is commonly used to specify movement of objects, a large amount of description must be specified. This problem is very serious and requires automatic animation to solve it. To automatic animation, operation must be generated automatically according to the environment.

Database

Shape and movement are used only once and cannot be applied for other animations. This means that data must be generated from scratch for each new production. A general-purpose database or related technology will help allow data to be reused.

Fujitsu has developed FUSION, an integrated computer graphics environment based on a high-level interface and object-oriented framework. In section 3, we will describe on FUSION.

RAY TRACING

Background

Recent CAD systems such as those used in industrial design have required the generation of high-quality images. Optical effects such as reflection, refraction, and shadow have become very important in the modeling. Although ray tracing can generate images of excellent quality, it suffers from lengthy computation times that limit its practical use. The image generation time of conventional ray tracing may take several hours[1].

A number of studies have been proposed to accelerate ray tracing. One group of studies uses coherence [2] to reduce computation time, especially for ray-object intersection calculations, which is the most computationally intensive part of ray tracing. Space partition techniques use object-space coherence that capitalizes on the relative spatial position of the objects. There are two types of space partition schemes, octree and voxel. We[3] and Glassner[4] presented the algorithm that traverses octree data structures. Although these method improved processing time, the speed is still less than ten times faster than conventional ray tracing. There are two reasons why the octree scheme did not cause much improvement. First, the overhead for traversal of the octree data requires extra computation in addition to the traversal itself. Backtracking is necessary when traversal fails in one child node. Second, since the space is dynamically divided when the octree is built, the size of the octant (cell of octree) can not be known except for the current one. It does not allow the algorithm to incrementally calculate the next voxel element. For these reasons, we chose the voxel data structure as the representation in the space partition scheme[5],[6],[7]. we will present a voxel traversal algorithm that efficient determines the next voxel element for the ray to proceed.

Another group of studies involves the parallel computation of the ray tracing algorithm. The parallelization of ray tracing capitalizes on the computational independency of individual rays that pass through the different screen pixels. Load balancing must be considered so that performance increases with the number of cells. In Links [8] the task of each processor is allocated dynamically during the processing. When a processor finishes its work, the host processor reallocates a task to it. The problem with this method is that it suffers from overhead due to communication. We propose two static load distribution schemes that allow the load to be balanced between the processors without overhead due to communication. In section 2.3, the implementation of the parallel ray tracing algorithm on the CAP that we developed is discussed.

While the number of techniques described above have improved the performance, they are still computationally too expensive for interactive use. Those studies based on object and ray coherence, were primarily devoted to speeding up the generation of still images. Although the methods from these studies can speed up image generation for dynamic sequences, another approach that capitalizes on frame coherence is more effective for interactive and successive image generation.

We think that the designer improves the model and the images incrementally until he is satisfied with the image. In a succession of incremental modeling phases, the designer refines parts of the image, for instance, he might change a color, and then move an object. When little change is made between frames, overall ray tracing for each frame is unnecessary. For this incremental design process, a partial update method for ray tracing is required that recomputes only the pixels that change between successive frames.

In image-space algorithms, the relationship between object changes and parts of the image to be recomputed are rather simple. Ray tracing based on global illumination model in the object space must handle reflections and refractions. As the object moves, its image on a reflecting object (for example a mirror) must be changed. As far as we know, no utilization of frame coherence in ray tracing has been made.

We developed a new approach [9],[10] to the problem in section 2.4, which enables ray tracing to generate dynamic sequences of images with little additional computation using the frame description of the previous image. We call the method *incremental ray tracing* since it allows the ray-traced images to be updated by partial recomputations during dynamic sequences. It deals with changes in the model, such as transformation, scaling, and addition or deletion to the database, and changes in surface properties such as color, reflectance, and mapping in a unified manner. Changes in shadows caused by the above changes and the movement of the light source are also included. The intersection tree has been extended to include the intersection point, surface normal, and related data that involves the path through which the ray propagated using voxel partition scheme. Because these description depends on the viewer position, the method works for a fixed viewpoint. In spite of the limitation caused by the view dependent description, the high-speed and high-quality image generation for successive frames that this method provides is useful. The result shown in section 2.5 demonstrates performance improvements as compared to conventional ray tracing.

Traversal algorithm

Concept of the space partition method In conventional ray tracing, each object in the environment is tested for intersection. Most of these tests useless because only a few objects intersect with a given ray. The space partition approach is more efficient since intersection calculations are done only for the object candidates.

The key idea behind the space partition method can be considered as *divide and conquer*. Intersection calculation time is significantly reduced by subdividing the enviror-ment into small orthogonal volumes, (voxel) and performing intersection calculations for only the objects that are likely to intersect with the ray. For each voxel object identifiers are recoded if the objects exist within the voxel. As a ray propagates from one voxel to the next, the objects in each voxel become candidates for ray-intersection calculation. Rejecting the objects that are not likely to intersect with the ray, the algorithm avoids many ray-object intersection calculations. The method can be applied to not only primary rays (the original rays from the view point) but also to shadows and reflected and refracted rays.

Voxel traversal algorithm The movement operation that finds the next voxel element for the ray to proceed must be fast so that its calculation does not cause additional overhead. We present a new traversal algorithm that efficiently determines the next voxel element for the ray to enter from the current voxel element in an efficient way. The method is called *the Parametric 3DDDA (3 Dimensional Digital Differential Analyzer)* since it uses the parameters of the ray (t in $\mathbf{X} = \alpha t + \beta$). The DDA algorithm [11] was originally used to obtain a rasterized straight line in two dimensions, where the successive pixels along the line are efficiently determined using only incremental addition. This concept can be applied to the traversal algorithm over the voxel data structure. The 3DDDA algorithm determines the next voxel element by using only incremental addition as well as the original DDA algorithm, although the 3DDDA algorithm performs in three dimensions rather than two.

The key idea behind this algorithm is to retain the points on the boundary planes of the voxel element. These points are then used to guide the next voxel element for the ray to proceed. We refer to points on the boundary along the ray with the parameter t. The value of t increases as we move away from the origin, where t is 0. Three t parameters, t_x,t_y, and t_z that indicate the points X_x,X_y, and X_z on the boundary with which the ray will intersect, are provided as shown in equation 1.

$$X_i = \alpha_i \cdot t_i + \beta_i \qquad (i=x,y,z) \qquad (1)$$

The selection which component of the voxel index changes is based on the following observation. Three t parameters predicate the next voxel element with respect to the current one. The minimum value out of three causes the component to change. Figure 1 illustrates this in two dimensions. Parameters, t_x and t_y indicate the points on the boundary where the ray will intersected. When t_x is smaller than t_y, the ray traverses the boundary of x ahead of y. This means that the x component of the voxel index changes by the value +1 or -1 according to the sign of the direction of the ray (α_x). In this manner, the determination of component to changes is made by only examining the minimum value of the t parameters.

After determining the minimum value, the parameter has to be updated for the next evaluation. The precomputed step, Δ_x,Δ_y, and Δ_z that represents the distance between the boundary planes along the ray is added to the t parameter which is just selected. These precomputed values are determined as $\Delta x = \frac{unitsize\ of\ voxel}{\alpha_x}$.

The cost for one voxel movement requires one integer incremental addition for updating the voxel index, one floating incremental addition for updating the t parameter, and two floating comparisons for selecting the minimum value out of three values.

For an exception, the ray may not intersect with one of any boundary planes. For example, the ray with $\alpha_x = 0$ does not intersect YZ-planes of the voxel boundary. Setting the initial value $t_x = \infty$ can avoid the ray from intersecting with YZ-planes automatically because it never can be chosen as the minimum value.

Parallel ray tracing on CAP

CAP architecture Ray tracing algorithm is implemented on a parallel processor, *the CAP (Cellular Array Processor)* [12],[13], which we developed. The CAP is a MIMD-type parallel computer consisting of from 64 to 256 processors called cells configured in an 8 × 8 two dimensional array. Each cell consists of an i80186 MPU, an i8087 NDP, and 2M-bytes of local memory. Figure 2 shows the CAP system. Every cell is connected by two buses. The Command Bus is used to download data and programs from the host computer. It can also be used for broadcast communication from the host computer to the cells or from a particular cell to the host computer or other cells. Each cell is connected to four neighboring cells to enable local communication. Cells at the boundaries are connected to cells at the opposite sides, so the network topology resembles be seen the surface of a torus.

For the CAP's graphical facility each cell has video memory within it. Images in video memory in cells are displayed on a monitor through the Video Bus. A significant feature of CAP's capability for display is a variety of screen partitioning modes, in which any cell can output its results, usually an intensity value, at any pixel on the screen. Each cell knows its identifier and the pixel location on the screen that the cell handles. Optimal load balancing of ray tracing can be achieved with this facility. The task is distributed to many cells and the corresponding image that the cell produces is displayed.

Load balancing The ray tracing algorithm is analyzed to consider the parallel ray tracing. Since the ray is traced backward from the view point to the light source, the computation for the rays that pass through the different pixels can be done independently. The parallel ray tracing algorithm can capitalize on the above computational independency of the individual ray. In parallel ray tracing each processor has a copy of the object database and processes rays that pass through the pixel element assigned to the cell in parallel.

Load balance must be considered so that performance increases with the number of cells. Our method uses static load distribution to allocate part of the screen to each cell instead of dynamic load distribution that uses communication. In the dynamic load distribution method, overhead may become too expensive when a large number of processors are used, since amount of reallocation and communication increases with the number of processors. The static load distribution algorithm allows the load to be balanced without communication.

We evaluates two modes of static load distribution method, the line mode and the dot mode. The mark of each Figure 3 shows the pixels which are processed by a processor. In the line-mode, cells are allocated each 63th scan line. Cell-0 handles the 0th scanline, the 65th scanline, and so on. Cell-1 handles the first scanline, the 65th scanline, and so on. In the dot mode, each cell performs ray tracing on every eighth pixel horizontally and vertically. Cell-0 handles the top left pixel of the screen. Next it handles the eighth pixel to the right. After completing the scan, the process is repeated eight pixels lower. Cell-1 process the right neighboring pixels that cell-0 handles. In this way, the whole screen is processed by the 64 cells.

Incremental Ray Tracing

Outline of algorithm The method has two stages (Figure 4). In the first stage, standard ray tracing is performed to generate an initial image for the successive frame sequence and to construct the data structures. These descriptions called the history are then used to quickly update the image during the second stage, which involves an interactive updation process. The partial recomputation due to a change of geometry or of surface property parameters are performed in the second stage. The frame description is maintained for subsequent frames. As long as the designer does not change the viewing parameters, this loop continues.

Before continuing the discussion, we will provide the terminology used in this paper. A *ray-set* is defined as all the branched rays of a given ray. We say a ray change occurs when the intensity assigned to the ray changes, which is caused by a change of the visible point or of the surface properties. The intensity of each pixel associated with a ray-set must be updated when one of the rays within the ray-set is changed. For clarity we will discuss the method for changes in geometry and changes in surface properties separately, although they are unified later.

The flow of the partial update algorithm is described below. First the *tree traverser* searches for the ray change by traversing the intersection tree in a top-down fashion. During the traversal, the test whether a given ray changes is done by the *detection algorithm.* It examines the change with the *voxel traversal history* and the *hash index* which are described later. The *detection algorithm* must work as fast as possible to obtain good performance. When a ray is expected to change, its visible point is recomputed. The new visible point can be determined using the history information that limits the intersection recalculation. If the visible point has changed, then shading recalculation is done to determine the new intensity at the surface. Otherwise the saved intensity value in the frame description is retrieved and assigned to the ray. If the change is detected, then tree traverser propagates the updated intensity to the pixel value.

The key idea behind the method is to localize the influence of a changed object with a voxel partition. With the description that contains the path through which a ray propagated in the previous frame and the voxel data structure that contains a flag that indicates whether any object within the voxel has changed, the intersection recalculations can be limited. Testing if the ray is changed by the changed object is done efficiently with a hash index that represents the ray's path. Also, an optimal traversal is required to avoid an exhaust search and to minimize the access cost of the data structures.

History data structure Figure 5 illustrates the data structures that contain all of the information necessary to partially update the ray tracing. This section describes these data structures.

(1) Extended intersection tree

For each pixel a intersection tree is constructed. The node is extended to include the following information in addition to the intensity value.

- ray parameter, such as direction and start point.
- information on the visible object, such as its visible point, normal vector at the visible point, and index of the surface properties.
- pointer to the voxel traversal history data structure described below.
- hash index (described later).

(2) Voxel traversal history

Our ray tracing system reduces the number of intersection calculations to speed up image generation by partioning the environment with a voxel data structure. The *voxel traversal history* contains the voxel index sequence through which the ray of the node propagated in the previous frame. Loosely speaking, voxel traversal history is a rough representation of the trajectory of the ray. For each ray this list is constructed and linked to the ray's node of the extended intersection tree.

(3) Intersection history

Intersections with objects within each voxel element are saved in the *intersection history*, which is linked to the corresponding list element in the voxel traversal history. Each record of the intersection history contains an object number and the relative depth of the intersection from the start point of the ray. Intersection recalculation during the partial update can be reduced with this information. Recall that the visible point is the closest point between the intersection points and the ray. The intersection points for unchanged objects are retrieved and used for the candidates of the visible point. In Figure 6, suppose the object that was the visible object moves, then the object previously obscured is revealed. In this case, intersection recalculation for the unchanged object is not necessary.

(4) Other data structures

- The voxel data structure is constructed in a fashion similar to previous works. The difference is that each voxel element has a flag to represent whether a change has occurred within its volume in addition to the object records.
- The object database contains information on the size, translation matrices, indices of surface properties tables, and change flags of the objects in the environment.

<u>Detection of changed rays</u> One possible choice for recalculating the visible point is a simple algorithm. It computes the intersections between every ray and every changing object in an inefficient manner. The computational cost is O(NR) where N and R denote the number of changing objects and rays, respectively. This method would be slow especially for complex scenes because R represents the complexity. Optimally the computational cost should be proportional to the number of the changing objects.

We will present new algorithm which is efficient since the effect of such changes can be localized. Voxel encoding is employed to represent each ray path in the previous frame of the image and to localize the influence of the changes in object geometry. Figure 7 shows a top view of the localized regions with the voxel partitioning scheme.

To begin every second stage (partial update), the voxel elements in which the changed object exist are found and flagged. Also, the object database corresponding to the updated object is flagged. The visible point of a ray may have changed in any of the the flagged voxel elements along the ray path that is encoded and saved in voxel traversal history. This traversal is efficient because it only traverses the voxel through which the ray passed in the previous frame. Note that the visible point in the changed voxel element is not always changed because the resolution of voxel partition is rough, currently 32 voxels along each axis.

Since we have assumed that the number of the changed objects is small and changed rays are few, it is inefficient to examine the entire voxel traversal history to detect changed rays. We will present a more efficient algorithm that exploits a hashing technique to rapidly test whether a given ray changes. Hashing is commonly used as a searching technique for the symbol table in compilers [14]. A hash function which takes the sum for example, is applied to the set of voxel indices. The resultant hash value called the *hash index* is recorded in the node corresponding to each ray. The hash index can be thought of as a concise representation of the path through which the ray propagated. A flag in the hash table is set if the changed voxel index is a member of the voxel index set corresponding to the hash index (Figure 8). Each frame is preprocessed to set the flags by scanning the voxel index set of every hash index. During the update, the flag of the hash index of a given ray is accessed to test if the ray changes. Since testing can be made by examing the flag in the hash table, which requires only one comparison instead of scanning every voxel element, the hash index reduces the computational cost. An additional merit is that the number of hash indices is much smaller than the number of the total rays because of ray coherence. It allows little amount of data space.

Figure 8 illustrates how to construct and use the hash index. The voxel index set of a ray is obtained by sorting the voxel index sequence through which the ray propagated. Sorting is done for two reasons. First, only unique combinations rather than the ordered sequence is needed here. Second, the sorting reduces the size of the hash table, which allows efficient construction and access. In this example, the sorted index set (5 6 10 11 15 16 20 21) is obtained from the ordered voxel index sequence (6 5 11 10 16 15 21 20). This index set is then recorded in field corresponding to hash index H2 that is sum of the voxel index set. Hash index H2 is assigned to the ray, *ray-3*. If a change occurs within the voxel element indexed 10, then preprocessing will set the flag of hash index H2. During the update, the flag is accessed with hash index H2 to test whether the ray *ray-3* propagated in the voxel element in which the change occurred.

<u>Recalculation of visible point</u> The recalculation of a visible point is performed when the algorithm, proceeding along a ray's voxel traversal history, comes to the changed voxel. The new visible point can be determined from the ray's intersections with the unchanged objects and the newly calculated intersections with the changed objects. Intersections for the unchanged objects are retrieved from the intersection history, and become candidates to be the visible point. Also, the recomputed intersections for changed object are added to these candidates. The visible point can be determined by evaluating the distance from the start point of the ray. Since the intersections for the unchanged objects need not be found, this algorithm can reduce the number of ray-object intersection recalculations, which is the most computationally intensive part of ray tracing.

<u>Changes in surface properties</u> Changes in surface properties can be performed just as for the branched ray. Since only the intensity value and the normal (if a bump mapping is applied) are changed, the structure of the frame description need not be changed. The test of whether the surface properties at a visible point has changed is made by matching the surface properties index of the visible object with the specified index of the surface properties. Traversal, recomputation of shading, and the propagation of the intensity is the same except that visible point recalculation is unnecessary. In stead of a hash index, a flag is kept for each surface property index and is used for the selective search. The position in the list of flags of a surface property index is set to 1 if the index is present in

any of the branched rays of the current ray. With both the flag and the hash index, the two methods can be unified easily.

Implementation This algorithm is implemented on the *CAP*. The management of the large amount of data used by our method is achieved using the distributed local memory of the CAP system. The total memory size is 128MB. It provides efficient access so that very fast performance is obtained. Data distribution also must be taken into account to prevent huge amount of history data from being assigned to a particular cell, which would exceed the 2MB memory of each cell. We can be assured that this will not happen by the following consideration. The amount of data is proportional to the number of rays, which is also proportional to the processing time. As shown above, the processing time in dot mode is balanced. Consequently, the amount of data is balanced.

Data is maintained for the next frames during the second stages. Since this history data structure may change, such as through update, construction, and deletion, this data structure is implemented as a heap.

Although a large data space (approximately 50 MB in total) is required, the amount of actually accessed data can be small. The traversal algorithm and the detection algorithm limit the access cost over the data structure by ignoring rays that are not likely to change using the global hash indices. For example, using the global hash index of the primary ray, traversal of any nodes of the tree is omitted when the global hash indices' flag is unmarked. The next section shows this effect.

Results and discussion

The results of the voxel partition scheme Figure 9-a shows the model SPHERES with 125 spheres used to demonstrate the performance of the space partition method. The test image was calculated at a resolution of 512 × 384 pixels. No antialiasing was performed for the tests. Table 1 shows the processing statistics with respect to the number of the partitions along each axis. This method reduces the computation time by a significant amount in comparison with conventional ray tracing that does not use any accelerating technique. Note that performance of conventional ray tracing is equal to the performance of one voxel partition in Table 1. For the model SPHERES with 6858 spheres, this method is approximately two hundreds times faster. While the number of partitions is small, the performance increases significantly. The processing time decreases from 13094 seconds to 2856 seconds when the number of partitions along each axis changes from 1 to 2. However, as the number of partitions increases the amount of performance increase becomes less due to the additional overhead for traversing the voxels. If the number of partitions becomes large enough, processing time actually increases. Note that the optimal number of space partitions depends on the model and the number of objects. Comparing the model with 16 spheres and 6858 spheres, the optimal voxel resolution of the model with 16 spheres is 8, while that of 6858's is 32.

Table 1. Processing time vs. number of voxel partitions.

Number of objects	1	2	4	8	16	32	48
16	102	39	23	20	21	25	30
512	825	212	68	34	28	31	36
1000	1537	450	126	59	38	38	43
4096	13094	2856	678	211	104	96	101
6858	32621	7306	1557	482	219	162	163

The model is SPHERES. Time is in seconds.

Load balancing Table 2 lists the statistics obtained from the line and the dot distribution modes. The difference ratio is defined as the ratio of difference between the maximum and minimum processing time compared with the maximum processing time. The difference ratio in the dot mode is within a few percent while that of the line distribution mode ranges more than ten percent. This implies that the dot mode distribution scheme works effectively. Figure 10 shows how performance increases with the number of cells for both dot and line load distribution modes represented the dotted and the solid lines. The graph is normalized by the processing time of 16 cells. In the dot mode, the performance factor of 64 cells compared with that of 16 cells is 3.96, which is almost proportional to the number of cells. On the other hand, that of the line mode is 3.65. This implies that performance in the dot mode is expected to increases for a large number of processors.

Table 2. Load balance statistics

Model	Line mode min - max (sec) difference ratio(%)	Dot mode min - max (sec) difference ratio(%)
SPHERES with 125 sphere	84 - 93 9.7%	88 - 86 2.3%
Chess	259 - 324 20.0%	281 - 294 4.4%
CAD	637 - 792 19.6%	667 - 691 3.5%

(*) difference ratio = (max - min) / max × 100

Comparison with large scale computers To demonstrate the performance and effectiveness of the ray tracing algorithm on CAP, we compare the performance of the CAP to that of a large scale computer, *Fujitsu M-380 (15 MISP)*. The processing time is measured on the CAP and M380 for the same algorithm. Table 3 lists the processing time for the models shown in Figure 9, which demonstrates that the performance of the CAP is comparable with the M380. Furthermore, ray tracing on the CAP has several advantages over a large scale computer. First, cpu time is equal to the elapsed time in dedicated machines such as the CAP, while the ellapsed time is usually much longer than the cpu time in a large scale computer. Since interactive image generation requires high-speed processing this characteristic becomes very important. Second, the performance increases with the number of processors in the CAP. Much higher performance is achieved when a large number of processors are used.

Table 3. Performance comparison of CAP(64 processors) and M380

Model (number of objects)	CAP (sec)	M380 (sec)
SPHERES (125)	72	60
CHESS (144)	155	128
CAD (178)	20	20
PISTON (179)	40	33

The result of the incremental ray tracing For each ray-traced image, several changes are made. We have provided the processing statistics for the entire ray tracing and the partial update shown in Table 4. All computing times are listed in seconds. The test images were calculated at a resolution of 512 × 384 pixels. No antialiasing was performed for the tests. Although approximately 100 percent more time is required to construct the frame description in the first stage with respect to standard ray tracing, this method reduces the computation times for changes by a significant amount in comparison with only using standard ray tracing. The relative improvement for a change of color is around 40 times faster, for an object change around 10 times faster, and for shadow changes due to the light source movement around 4 times faster.

Table 4. Timing test of incremental ray tracing.

Model	Chess without shadow	Chess with shadow	Spheres without shadow
entire	70.4	101.4	113.3
change-1	3.3 (21.2)	5.8 (17.4)	6.7 (17.0)
change-2	7.2 (9.9)	13.8 (7.4)	9.4 (12.1)
size	4.1 (17.3)	6.9 (14.7)	8.2 (13.8)
light	7.6 (9.3)	33.8 (3.0)	24.3 (4.7)
color	0.8 (83.8)	0.9 (119.3)	1.5 (77.1)

Processing times are listed in seconds. The relative performance is listed in parenthesis, which indicates the ratio with respect to the entire processing time.

FUSION: An Integrated Environment for Computer Graphics [15]

Motivation

Computer animation is difficult to make, especially for non computer specialist. It requires several types of processing at one time. These processing should be integrated into a single environment. A multi-window system [16],[17] allows several processing to run concurrently. With proper connection between these processes, it is possible to build very flexible user interface. New interfaces which can manage non-linear data structures, called *hypertext*, was reported in [18]. These systems allow the user to interact directly with chunks of data and to establish relationships between data items by linking them.

FUSION is a highly interactive system that supports to make object model and animation. It integrates several types of processing into unified environment based on a multi-window system. To do this, we propose a structured window system. In this system, the user can define a relationships between windows. Based on a structured window system, interactive modeling system is implemented. Each system is disintegrated into modular components. One component corresponds to one window. By interactively connecting windows, the user can constructs a complex application. As a result, the user can easily constructs and modifies large scaled animation sequences.

Overview

FUSION is a window-based integrated environment for computer graphics which incorporates a set of tools for modeling, rendering, and making animation sequences into a unified framework. This system provides a design environment in which 3D animation sequences can be designed. FUSION was developed to achieve the following goals:

1. To provide an integrated environment in which several operations for computer graphics can be performed.
2. To offer a good user interface for non-computer specialists such as designers and animators.
3. To allow system designers to easily and rapidly construct high-level systems.

FUSION consists of the *FUSION/MODELER*, a geometric modeler, the *FUSION/ANIMATOR*, an interactive motion generator, and *FUSION/MMI*, a structured windowing system.

Each step is performed in a corresponding subsystem. These subsystems are interconnected by message passing communication and can run concurrently. The user constructs or modifies the application by creating forms and making interconnection for each form. The window manager creates and deletes individual forms. FUSION/MMI offers common user interface of forms and manages the intercommunication. The following section describes the subsystems in detail.

FUSION/MMI

The windowing interface we propose is called FUSION/MMI. One idea of linking windows is to use arcs. However, this causes an incomprehensible view for the user when several windows are connected with lines. We present a slot-fill representation to link two windows. Every window has slots, the slot size being equal to that of an icon. When two windows are being connected, the child window is converted to an icon and is placed in the slot of the parent window as shown in Figure 11. At that time, connection is established and the icon reverses its color to indicate this.

FUSION/MODELER

FUSION/MODELER is based on a simple concept. It consists of many forms. These forms on the screen are associated with primitive descriptors, CSG expressions and others. The user can model the geometry and appearance by writing the description on each form and linking them to construct a structure. The structured window can be used to represent networks. It thus has enough ability to describe CSG trees.

Forms are used to input direct manipulation with graphics tools, and textual description with panel tools. Since each form has an input format specific to its function, forms can be customized according to the task. For example, PrimitiveForm is used to specify the type and size of the primitive. Visual operations are used to select the type of primitive. Clicking a mouse button on the form displays the primitive type menu. Note that size-keywords such as height and radius change according to the type selected as shown in Figure 12.

Another interesting feature of the visual operation is that of describing CSG expressions shown in Figure 13. Each icon on the middle pane of the CSGForm represents a linked icon which represents an object. Clicking a mouse button between the icons displays a pop-up menu which contains boolean operators such as +,-,&(and), and the delete operation. These operators can be inserted between the icons and deleted with this menu.

AttributeForm shown in Figure 14 is used to specify parameters of surface properties. Since several parameters are required, the operation has to be simple and easy. AttributeForm is good example of panel facilities; the value can be specified with the slider. Coefficients for specular and reflection, which depends on the angle, can be specified by pointing the cursor to an arbitrary value on the graph and by interpolating between specified sample points.

In addition to visual operations on each form, linking is also easy to do since the images of the slots represent the allowable forms to be linked. A syntax error is reported immediately if a wrong link is attempted.

Forms corresponding to abstract objects In FUSION/MODELER, some of the forms can be considered as representing objects. For example, PrimitiveForm is the abstraction of primitive and CSGForm represents the group of objects which are connected by boolean operators. These forms have their own names and images. The image of PrimitiveForm represents the primitive's image itself and the image of CSGForm represents the image of the combined objects. This correspondence between objects and forms provides the user with an intuitive way of modeling.

Generic object To offer structured modeling, FUSION/MODELER introduces the object-oriented paradigm extended to support partof relationships. In this scheme, the user describes the class of object rather than making an instance itself.

Class can be considered as an abstract object whose features can be changed according to the parameter values. A wide variety of objects can be obtained from a class, which is considered as a parameterized generic object. Suppose we've defined the class description of chair which can generate a chair with four legs of a variety of sizes. By changing the parameter values for height_back, height_legs, and chair_color, a different chair can be generated. This mechanism differs from the scaling operation in direct manipulation since scaling operations can only change objects proportionally.

Describing the behavior of the object can also take advantage of parameterized objects. Procedural definitions of the objects and changing the joint value according to the time parameter is useful when making animations taking advantage of parameterized objects.

Partof relationship The partof relationship introduces the structure of a model. In some class definitions, other classes are used as parts. Reuse of the existing model is achieved and productivity of modeling should increase by using this scheme.

In the class definition, the user describes the parameters and their default values. In the scope of the class, the expression using the class parameters can be used to give the parameter to the part or primitive forms. Figure 15 shows Classform and PrimitiveForm which are for the Chair model. In this case, Class-chair has the height_back and height_leg parameters which enables those sizes to be changed independently. In addition, an expression such as (height_back - height_leg * 2) is allowed in the class scope. In this way, parameter values are propagated along the partof hierarchy.

FUSION/MODELER also provides an easy way of updating or modifying the model. Since a description of the model is preserved in the form, the description can be easily modified. The user can retrieve the saved form from the database by linking the DatabaseForm which indicates the model. Figure 16 illustrates the modifying process.

First, ClassForm appears on the screen containing the related forms. Clicking a mouse button on the desired slot shows the form linked to the class form, in this case the CSGForm. Changing the CSG operator from + to - redefines the objects easily.

Symbolic Positioning One cumbersome aspect of conventional modelers is related to the transformation description. We proposed the symbolic position [19] which represents the feature point of the object, and the scheme of translation with these symbolic positions. For example, we can reasonably call the top-point of the table as TOP. Suppose a cup is on the center of the table. It is natural to specify with the translation that the TOP of the table and BOTTOM of the cup should be connected. Since these symbolic points are abstract and independent to the size of objects, the user can implicitly specify the amount of transformation ignoring the specific size of the objects. Modeling at an abstract level is thus achieved.

We define the symbolic positions TOP, BOTTOM, RIGHT, LEFT, FRONT, REAR and CENTER for primitives. Symbolic positions for a class or user-defined objects are determined by the user. Any number of arbitrary names are available for symbolic positions of objects except for primitives. Any point of the object can be expressed: An expression of (TOP + RIGHT) points to the top-right point since a symbolic expression acts as a vector calculation. Another example is (TOP + RIGHT + [x y z]) which indicates the relative point [x y z] from the top-right point. Note that points other than the surface of the object can be expressed. The object name is placed before the symbolic expression to specify the object's point. An expression TABLE.(TOP + RIGHT) indicates the top-right point of the table.

Implementation

FUSION is implemented on a SUN workstation using SUNVIEW for the window tools. FUSION/MMI is responsible for the structured window system which performs visual operations and maintains relationships between forms with messages to send and receive. FUSION/MMI consists of two elements, forms and the database. In forms, FUSION/MMI manages a graphics representation and communication protocol between the forms and the database in Figure 17. Each form has its own processing to manipulate input, output, and resources and to perform computation. Each form is designed for use with a mouse and window items such as menus, buttons, and sliders. These items make forms easy to use.

Each form has only a part of the data related to itself. To guarantee the completeness, data should be managed by the database. Forms send their own data to the database via IPC (Inter Process Communication) when data is entered or modified. Forms are represented with a frame structure in the database with each frame corresponding to a form. Interconnection between forms is reflected as a frame relationship. When a form starts its processing, it sends messages to FUSION/MMI and a corresponding frame structure is created in the database. The user can interact with FUSION/MMI to create or delete linkage between different forms. Communication between forms and the database is implemented by typed, variable sized messages. To send a message, data is packed into textual format and passed by a datagram IPC packet. When a database or form receives an IPC packet, it reads the message and branches to the routine which specified the message type. Each form process can be made easily by using libraries which handle the panel interface and communication. The forms processes can be written in languages such as C, Lisp, or Fortran depending on the task. This allows the system developer various means to extend FUSION.

SUMMARY

We have developed the practical ray tracing algorithm using three techniques. First, the ray-object intersection calculation is reduced using the space partition technique and efficient traversal algorithm. It demonstrates that significant improvement is obtained, from several dozen to 200 times faster than that of conventional ray tracing. Second, this algorithm is implemented on parallel processor, the CAP. We proposed efficient static load balancing method. The dot mode static distribution algorithm demonstrates that load can be evenly balanced without communication. The performance increases linearally with the number of processors within a certain ranges. Third, we have presented a method for the partial updating of ray-traced images that is very useful for interactive modeling of geometry and surface properties for a fixed viewpoint. We capitalized on the geometrical properties of the fixed viewpoint and only slightly changing geometry to exploit frame coherence, so that only some of the image needs to be recomputed from frame to frame. Since there is less rendering to be done at each frame, the cost of ray tracing each frame is correspondingly diminished.

With those techniques ray tracing becomes practical for some computer graphics applications, such as CAD and animation which require high-quality images.

And also, we have described our approach for an integrated computer graphics environment which supports a convenient user interface for the designer. *FUSION* incorporates different processes into a single design environment. This integration takes advantage of a concurrent and structured user interface implemented on a structured window system. In addition, the object-oriented paradigm is appropriate for reuse and updating of the model. With this modeling scheme, large and complex objects are easily handled.

REFERENCE

1. T.Whitted, "An Improved Illumination Model for Shaded Display," *Comm. ACM* Vol.23, No.6, 1980, pp.343-349.
2. I. Sutherland, R. Sproull, and R. Shumacker, "A Characterization of Ten Hidden Surface Algorithms," *Computing Surveys,* 6(1), March 1974,1
3. Murakami, K., "Fast Ray Tracing using the octree partitioning," *27th Information Processing Conference Proceedings* October 1983,1537-1538,(In Japanese)
4. A. Glassner, "Space Subdivision for Fast Ray Tracing," *IEEE Computer Graphics and Applications,* 4(10),October 1984,15-22
5. K. Murakami, K. Hirota, "Ray Tracing System using the voxel partitioning on a Cellular Array Processor," *Graphics and CAD* 22(2),July 1986,1537-1538,(In Japanese)
6. K. Murakami, K. Hirota, and M. Ishii "Parallel Ray Tracing on a Cellular Array Processor," *IFIP : System Modelling and Optimization*, August 1987
7. A. Fujimoto, T. Tanaka and K. Iwata, "ARTS:Accelerated Ray Tracing System," *IEEE Computer Graphics and Applications,* 6(4),April 1986,16-26
8. Nishimura, H., Ohno, H., Kawata, T., Shirakawa, I., and Omura, K., "LINKS-1: A Parallel Pipelined Multimicrocomputer System for Image Creation," Proceedings of the 10th Symposium on Computer Architecture, SIGARCH(1983), pp.387-394.
9. K. Murakami and K. Hirota, "Incremental Ray Tracing on a Cellular Array Processor," *Graphics and CAD* 22(2),July 1987,(In Japanese)

10. K. Murakami, K. Hirota, and M. Ishii "Incremental Ray Tracing," *NCGA '88* 1988

11. J.E. Bresenham, "Algorithm for Computer Control of a Digital Plotter," *IBM System Journal, Vol.4,1965,25-30*

12. H.Ishihata, M.Kakimoto, "VLSI for the Cellular Array Processor," Proc. ICCD,October 1987,320-323

13. H. Sato, M. Ishii, et al. , "Fast Image Generation of Constructive Solid Geometry Using a Cellular Array Processor," *Computer Graphics,* 22(2),July 1985,95-102

14. V. Aho, and D. Ullman, "Principle of Compiler Design," *Addison-Wesley,* April 1979

15. K.Murakami and K.Hayashi, "FUSION: An integrated environment for Computer Graphics," *Graphics and CAD* ,July 1987,(In Japanese)

16. Paul E.Haeberli, "ConMan: A Visual Programming Language for Interactive Graphics", *Computer Graphics,* 22(4), August 1988, 103-111

17. Ranjit Makkuni, "A Gestural Representation of the Process of Composing Chinese Temple," *IEEE Computer Graphics and Applications,* 7(12), December 1987, 45-61

18. Jeff Conklin, "Hypertext: An Introduction and Survey," *IEEE Computer Graphics and Applications,* 7(9), September 1987, 17-41

19. K.Murakami and H.Matsumoto, "MEG: An Animation Language for Computer Graphics," *Graphics and CAD* 22(2),November 1986,(In Japanese)

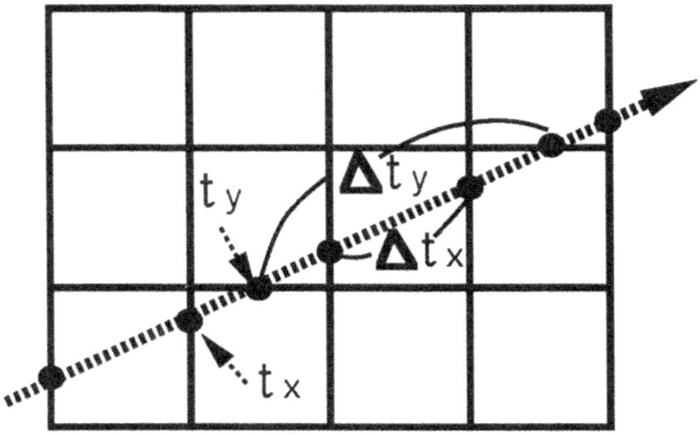

Figure 1. Scheme of voxel traverse.

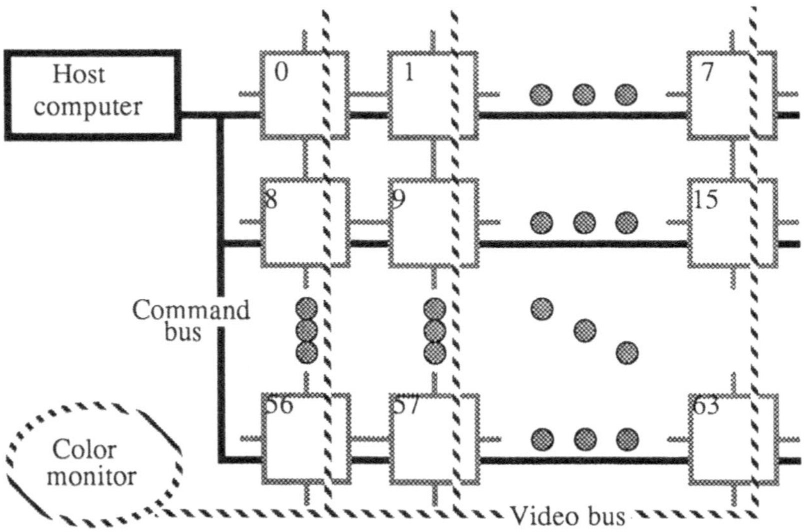

Figure 2. CAP Configuration.

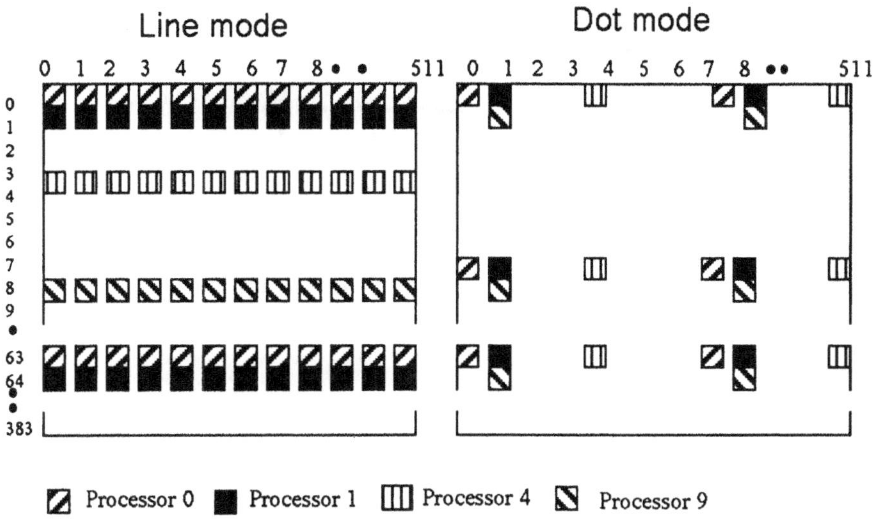

Figure 3. Line and dot distribution modes.

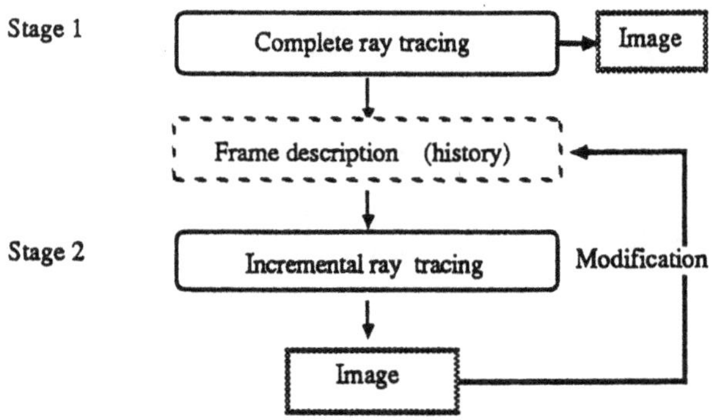

Figure 4. The two stages of image generation.

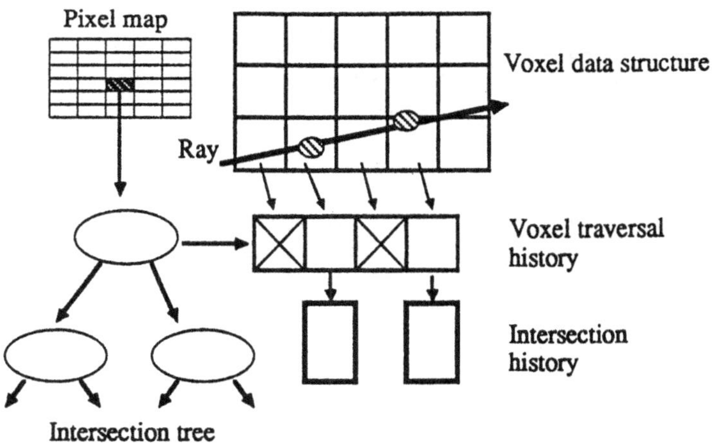

Figure 5. Data structure.

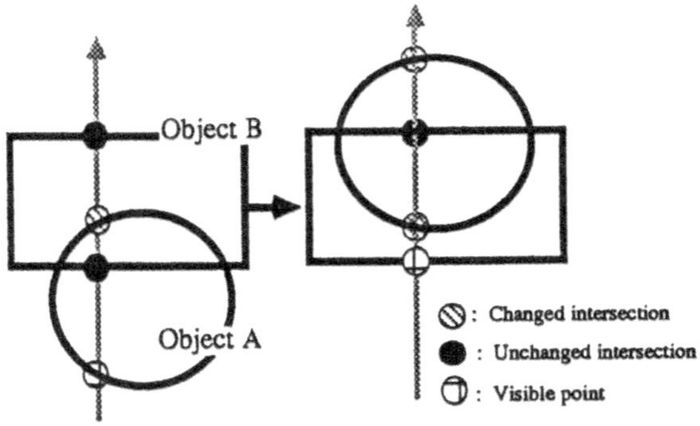

Figure 6. The use of the intersection history.

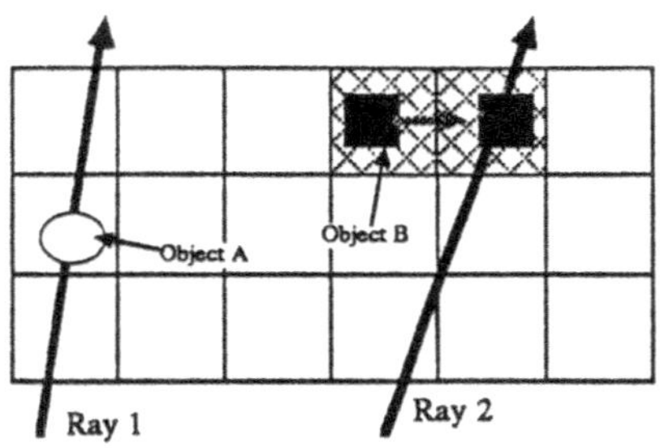

Figure 7. Localization of the influence of change.

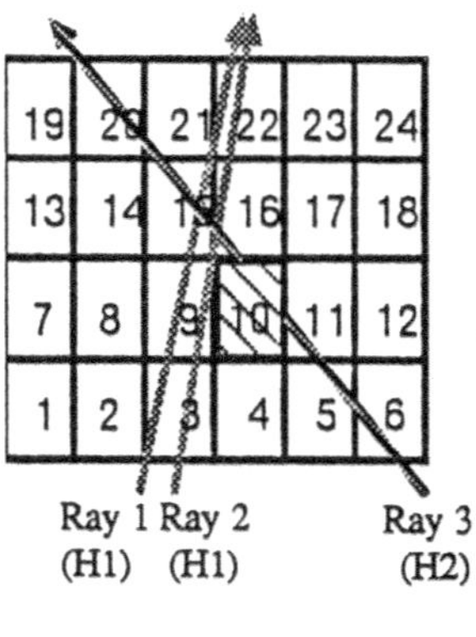

Index	Flag	Voxel index set
H1	Off	(3,9,15,16,22)
H2	On	(5,6,10,11,15,16,20,21)
~		~
	Off	

Figure 8. Hash table.

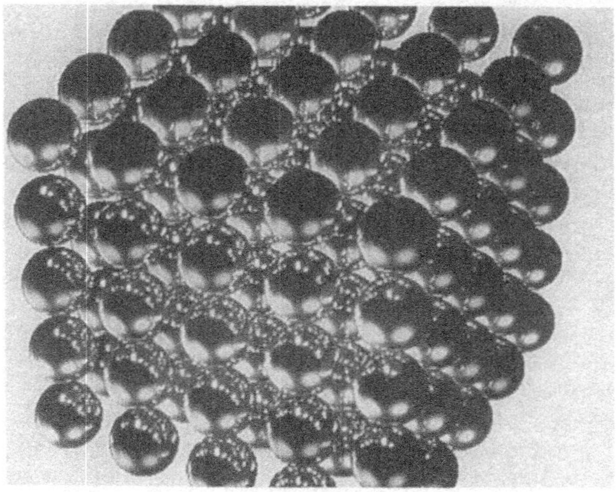

(a) SPHERES

(b) CHESS

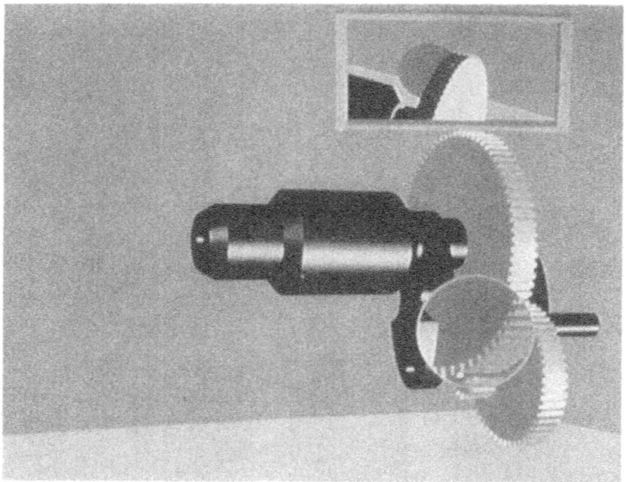

(c) CAD

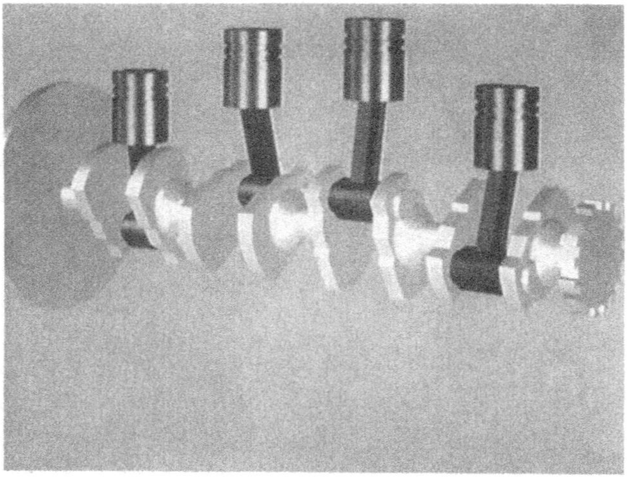

(d) PISTON

Figure 9. Models.

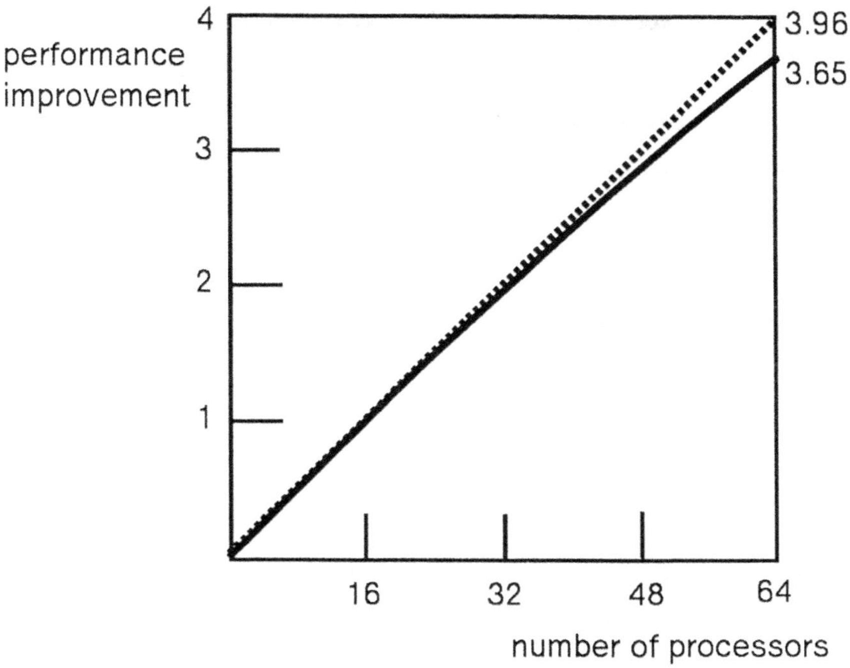

Figure 10. Performance with respect to the number of processors.

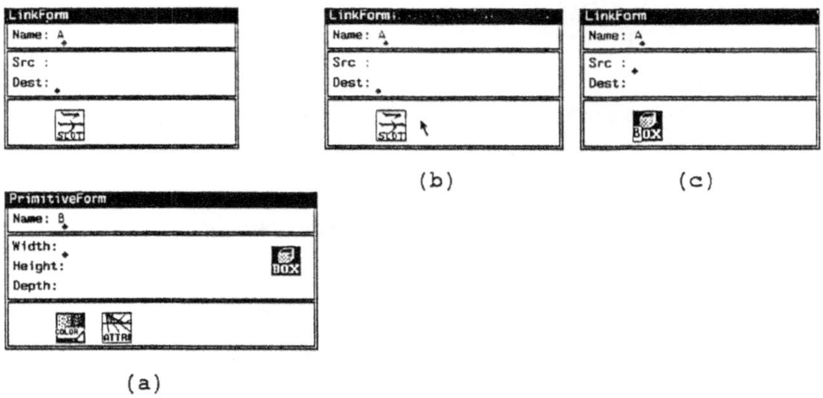

Figure 11. Connecting windows. The two windows in (a) are connected by first selecting link operation with mouse. In (b), the user specifies destination slot by mouse pointer. Then, window converts an icon and installs to slot shown in (c).

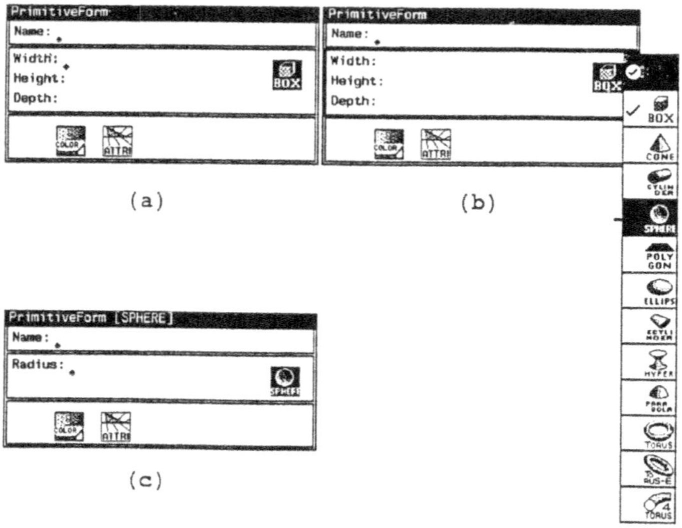

Figure 12. Primitive description. A BOX primitive is selected in (a). The size-keywords are width,height, and depth. New primitive SPHERE is selected from pop-up menu as show in (b). The size-keyword then changes to radius in (c).

Figure 13. Constructing CSG expressions. Each icon on the CSGForm represents the object. The CSG operator is specified by selecting appropriate operator as shown in (a). Then the selected operator is inserted before the icon in (b).

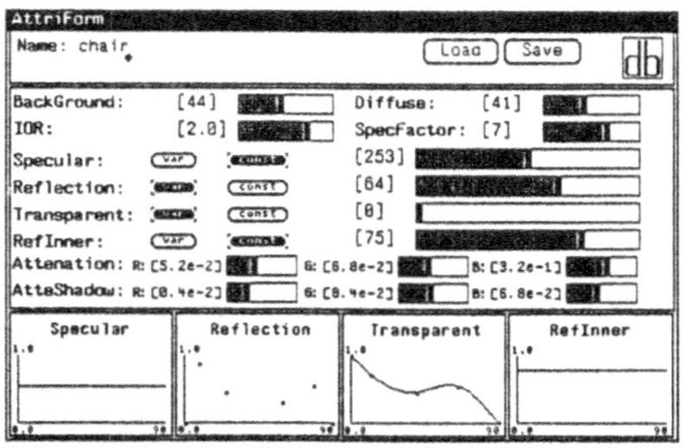

Figure 14. AttributeForm

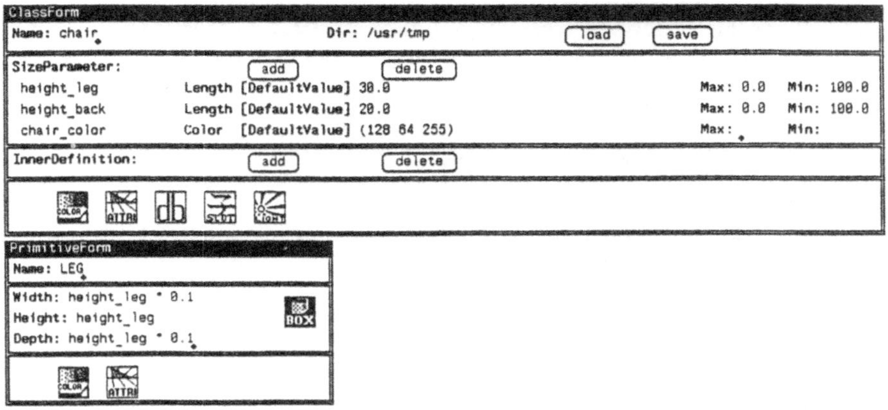

Figure 15. Class parameters. The class variables of class-chair is height_back, height_leg, and chair_color. These variables can be used for primitive parameters.

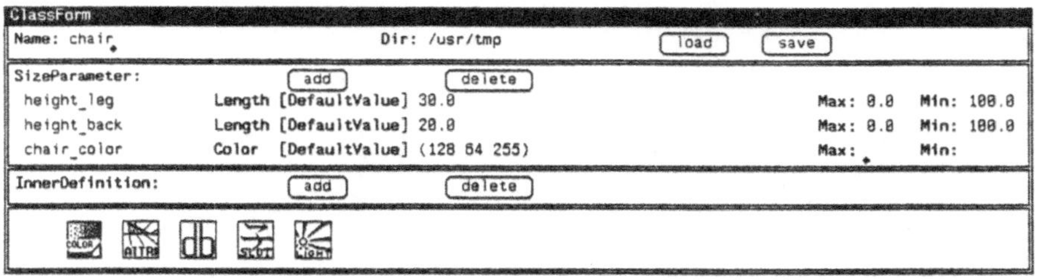

(a)

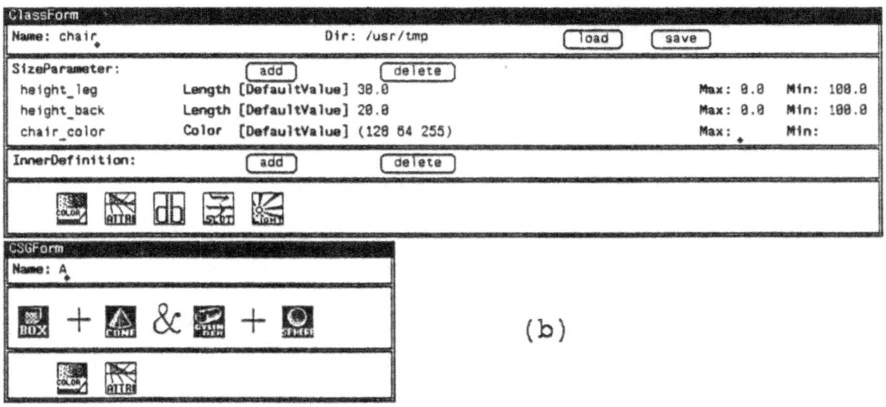

(b)

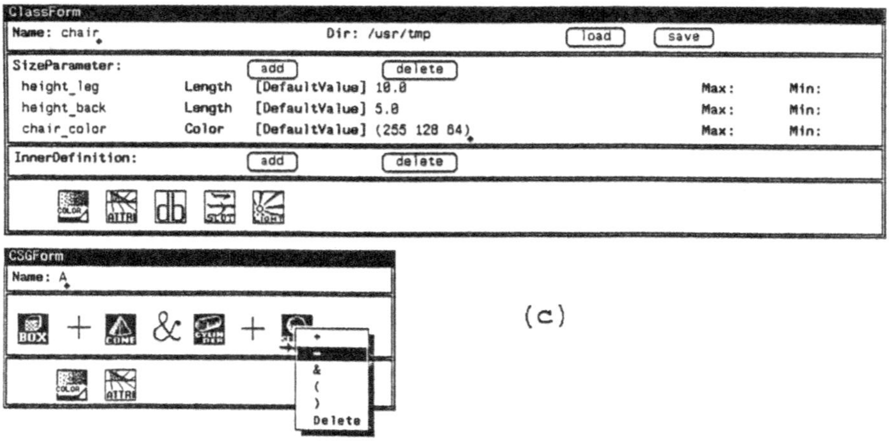

Figure 16. Modification of class. Class definition has the CSGForm as a child as show in (a). Clicking CSG slot causes the child form to appear in (b). It is then modified (c) to change + to -.

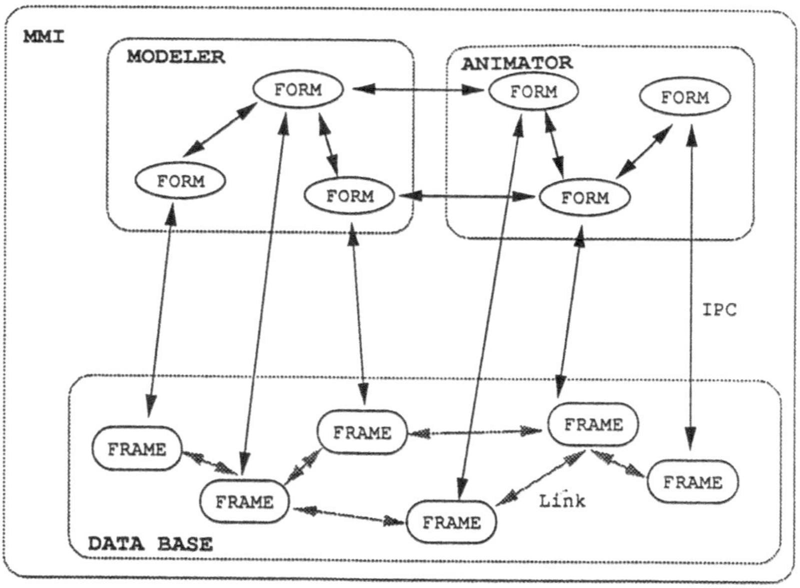

Figure 17. Implemetation overview. FUSION/MODELER and FUSION/ANIMATOR are built on FUSION/MMI. FUSION/MMI manages a graphics representation (form) and communication between the forms and the database. Each form is represented with a frame structure in the database. When the form is created or deleted, FUSION/MMI changes relationships between forms in the database.

Neurocomputers in Japan

H. Hayashi

Artificial Intelligence Laboratory, Fujitsu Laboratories Ltd., 1015 Kamikodanaka Nakahara-ku, Kawasaki 211, Japan

ABSTRACT

Neurocomputers designed after the human brain are "flexible" information processors composed of networks of many mutually connected neuron devices operated in parallel. Neurocomputers are capable of advanced parallel distributed processing and "learning". In recent years, research in the field of neurocomputers has extended into the fields of robot control, pattern recognition, and knowledge processing, with the construction of a flexible system as its goal. This paper outlines the research in field of the neurocomputers conducted in Japan, mainly that of Fujitsu.

NEUROCOMPUTER AND ITS CHARACTERISTICS

Figure 1 shows the mathematical model of a neuron. The neuron receives from other neurons, weights each input, and calculates a sum total. When the sum total is sufficiently greater than the threshold value, 1 is output. When it is sufficiently smaller than the threshold value, 0 is output. When the sum total is near the threshold, a value is output based on a sigmoid function that varies continuously from 0 to 1. Where, the weight and the threshold value are variable.

The neurocomputer is composed of many neuron devices and connected in a network as shown in Figure 1. Figure 2 is an example of a multi-layer neural network. For this network to function as desired, it must first go through learning process shown in the figure. In the learning process, the weight and threshold values are adjusted so that the output of the neural network matches a correct solution specified by a teacher.

The flexibility of a neurocomputer results primarily from the following two functions: (1) Learning function: Obtains and modifies a desired function by adjusting weight and threshold values. (2) Interpolation function: Outputs reasonably accurate values for points other than the learning sample points, using multiple analog input and output (see Figure 3).

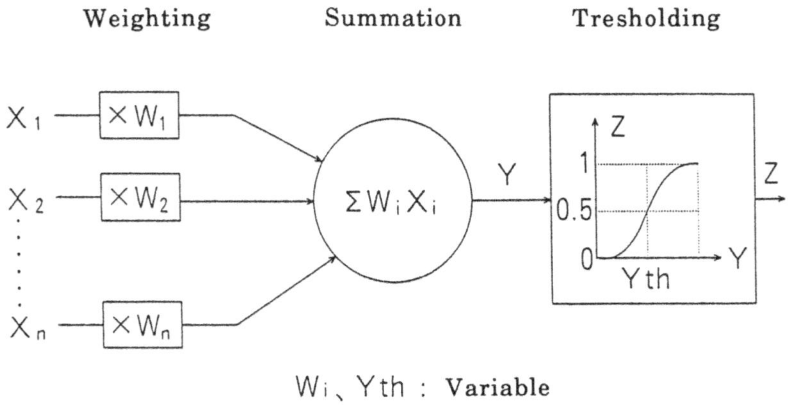

Figure 1. Mathematical model of a neuron.

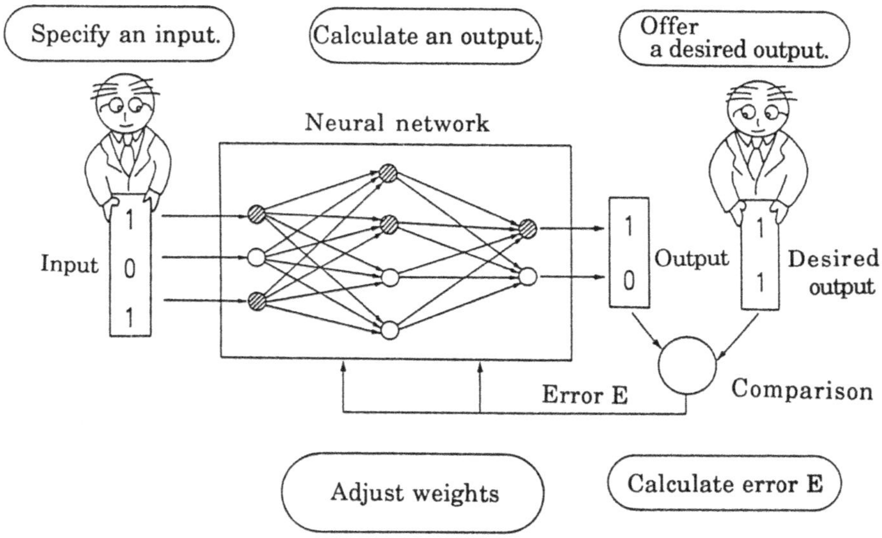

Figure 2. Neural network and learning.

RESEARCH ON NEUROCOMPUTERS AT FUJITSU

Research is being conducted on the neural network models and their learning methods, implementation tools such as software simulators and dedicated hardware, and applications of neurocomputing. This chapter introduces recent achievements.

<u>Neural network models and their learning methods</u>
Since most models currently under investigation require much time to learn a desired function, various high-speed learning methods are being researched. The following two high-speed learning methods have been

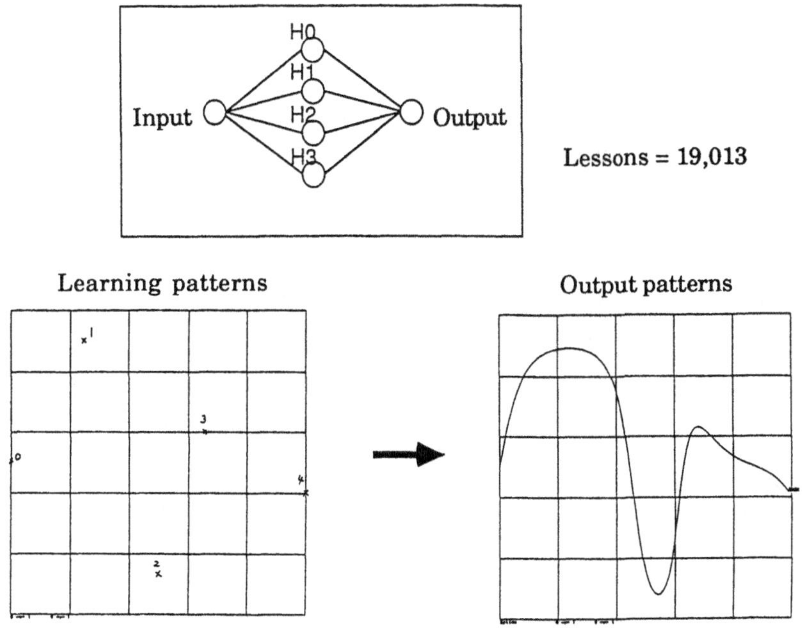

Figure 3. Experimental example of the interpolation function.

developed as of this date:

Pseudo impedance control method [1] Learning of the neural network is a connection weight adjustment for obtaining the desirable pattern from the output layer in response to a given input pattern. Rumelhart et al. [2] presented a method, called error back propagation method, for adjusting the connection weights in a layered network. The error, the difference of the desired and the current output, is fed back to adjust the weights between the layers through backward processing from the output layer to the input layer. As a result, the network learns the relationship between a given input pattern and the desirable output pattern so that adaptive data processing can be achieved. But there is a problem. This algorithm tends to stop the network at a local minimum.

To decrease the possibility of stopping at a local minimum, we proposed a new learning algorithm called the pseudo impedance control method which was derived through analogy with a mechanical vibration system, and handled network behavior as the following second-order forced oscillation system [3]:

$$J\frac{d^3w(t)}{dt^3} + M\frac{d^2w(t)}{dt^2} + D\frac{dw(t)}{dt} = -\frac{\partial E}{\partial w(t)}$$

$$\Delta w(t) = -\varepsilon\frac{\partial E}{\partial w(t)} + \alpha\Delta w(t-1) + \beta\Delta w(t-2)$$

$$\varepsilon = \frac{1}{J+M+D}, \quad \alpha = \frac{2J+M}{J+M+D},$$
$$\beta = -\frac{J}{J+M+D}$$

where J, M, D are positive constants, w the connection weight, E the sum of the squares of the errors, Δw weight change, and t the number of lessons (weight updates).

Figure 4 shows the necessary conditions of α and β where the network converges to the minimal point. We have simulated the various combination of ε, α, and β.

To investigate the convergence of the layered network, we tried to teach the network sine functions which are often used when accelerating or decelerating the robot arm. One period of the sine function was divided by 20 to get input data (see Figure 5). The formula used is given below.

$$Y_i = \frac{2}{5} sin(2\pi X_i) + \frac{1}{2}$$
$$X_i = \frac{i}{20} \quad \text{for } i = 0,1, \cdots 19$$

The experimental network is shown in Figure 6. Input pattern *Xi* is placed in the input layer; the output value is compared with teaching pattern *Yi* to teach the network. The network was taught 100,000 times.

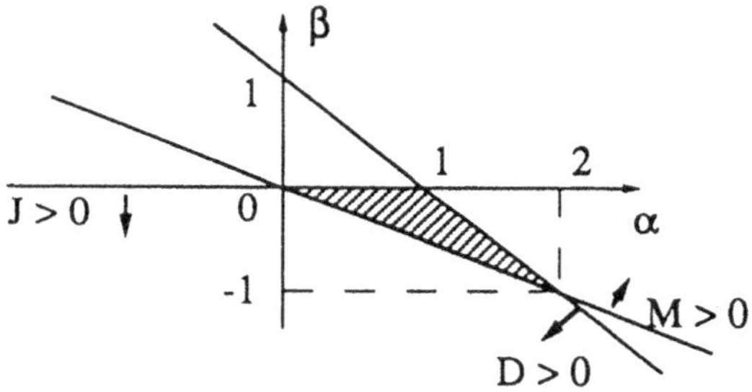

Figure 4. Necessary condition of α and β [1].

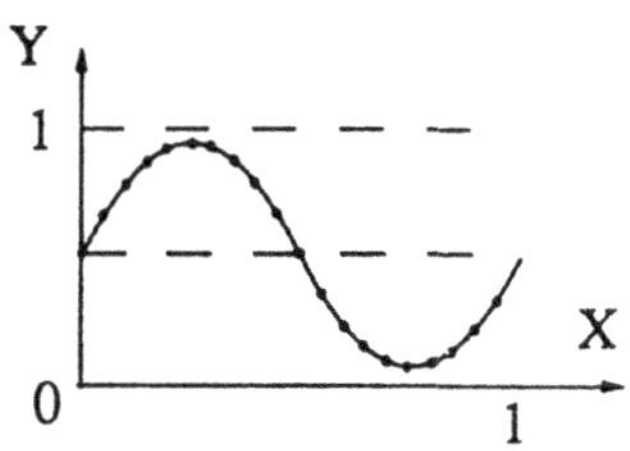

Figure 5. Learning patterns [1].

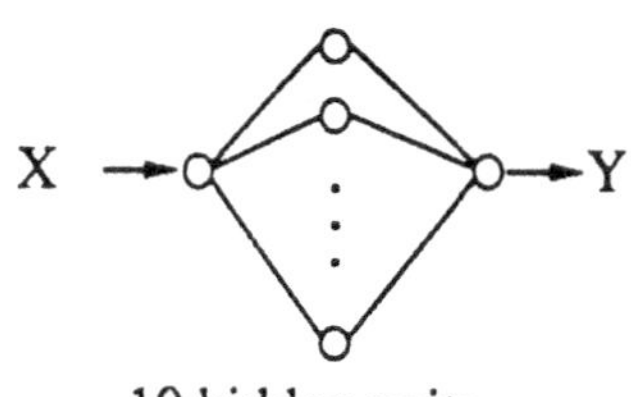

Figure 6. Experimental network [1].

Figure 7 shows the convergence in three cases: (1) α=0 and β=0, (2) α=0.9 and β=0, and (3) α=1.2 and β=-0.4, for ε=0.6 in each case. For quick convergence, ε should be as large as possible. However, case (1) does not converge because of a large oscillation. Case (2), the error back propagation method, converges because term α has a damping effect, but the convergence rate decreases gradually when the number of lessons increases; the error remains at a rather high level. This seems to occur because the network has been taught an analog signal represented by nonlinear function. There are thus many local minima, so the network cannot exit this high error state. When the term of β is added as shown in case (3), the final error is reduced although the process is oscillatory. It has now been revealed that the effect of β helps the network to exit the state local minima which are proved to be present in great quantity and to learn toward a minimum with a smaller error.

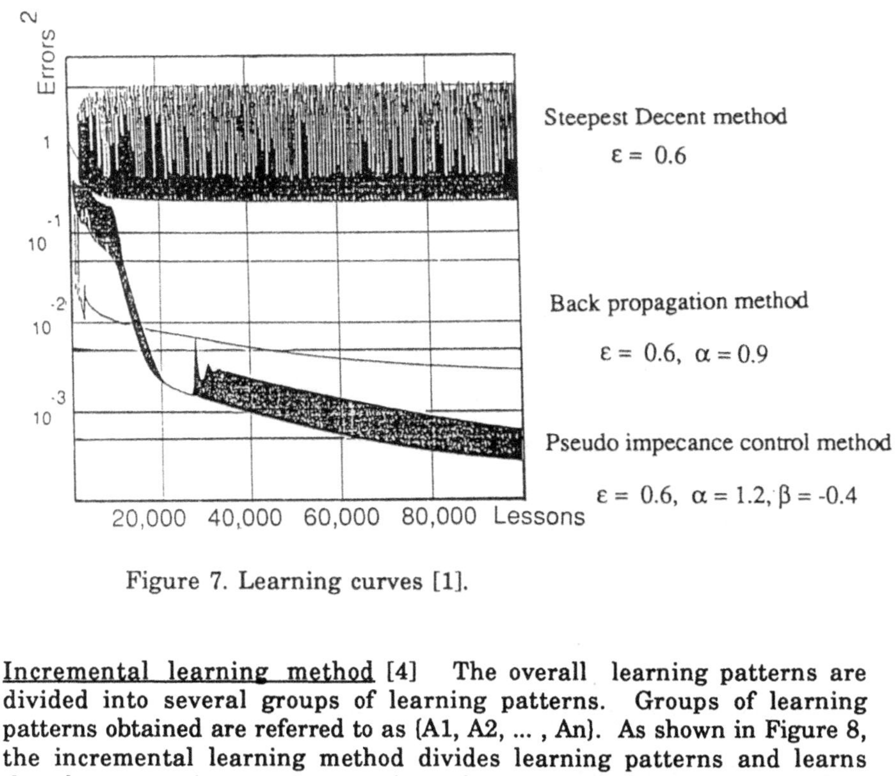

Figure 7. Learning curves [1].

<u>Incremental learning method</u> [4] The overall learning patterns are divided into several groups of learning patterns. Groups of learning patterns obtained are referred to as {A1, A2, ... , An}. As shown in Figure 8, the incremental learning method divides learning patterns and learns them by stages. A new group $\alpha 1$ = Ai+1 of patterns is taught to the network in which an overall group $\alpha 2$=A1UA2U...UAi of patterns has already learned. First, let the network learn $\alpha 2$. Then, let the network learn $\alpha 1$U$\alpha 2$ and fix the network in the state in which the learning of both $\alpha 1$ and $\alpha 2$ has been completed. In short, the incremental learning method lets the network each new group of patterns by going through a new learning phase and a review phase.

An experiment utilizing a layered network composed of a 64-unit input layer, a 15-unit hidden layer, and 26-unit output layer to recognize 8 x

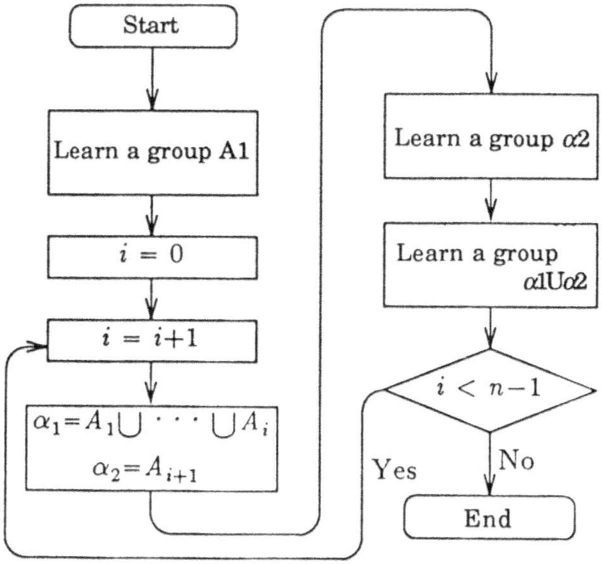

Figure 8. Incremental learning method [4].

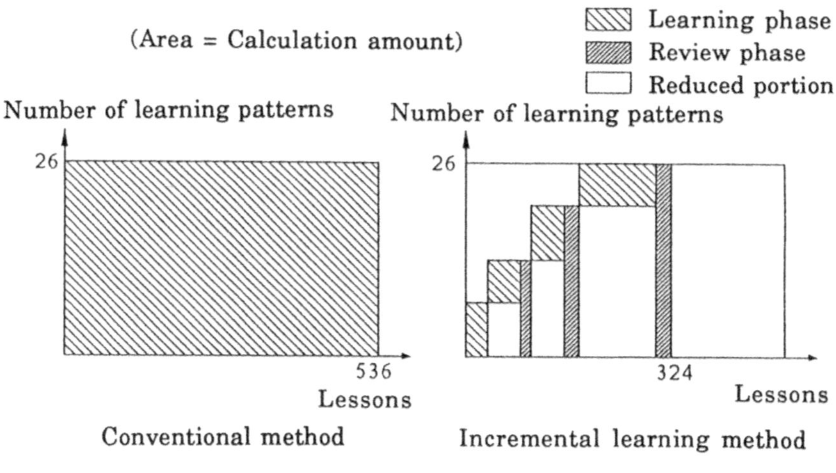

Figure 9. Experimental example of the incremental learning method [4].

8 dot alphabet characters was conducted. The following procedure shows how the characters were learned: (a) Divide patterns A to Z into four groups of patterns, α={A, B, C, D, E, F, G}, β={H, I, J, K, L, M}, γ={N, O, P, Q, R, S, T}, δ={U, V, W, X, Y, Z}. (b) Let the network learn pattern α. (c) Let the network learn pattern β. (d) Let the network learn patterns α and β collectively. (e) Let the network learn pattern γ. (f) Let the network learns α, β, and γ

collectively. (g) Let the network learn pattern δ. (h) Let the network learn patterns α, β, γ, and δ collectively.The learning speed of the ordinary learning method was compared with that of the above incremental learning method for networks in the same initial state. The incremental learning method is several times faster than the ordinary learning method, as shown in Figure 9.

Implementation tools

Software simulators Software tools have been developed that enable construction of a neurocomputer as described before on a conventional computer via software. These simulators run on workstations or personal computers, and model the operation of small-scale neural networks.

Dedicated hardware As the scale of a neural network increases, the software simulator cannot supply the necessary speed. For this reason, research and development of dedicated hardware to improve the execution and learning speeds being conducted. Figure 10 shows an example of an analog neuro chip configuration using semiconductor technology [5]. The analog neuro chip uses time division analog busses. Connection weights are kept in external storage, and can easily be varied to prevent the number of wiring connections from increasing.

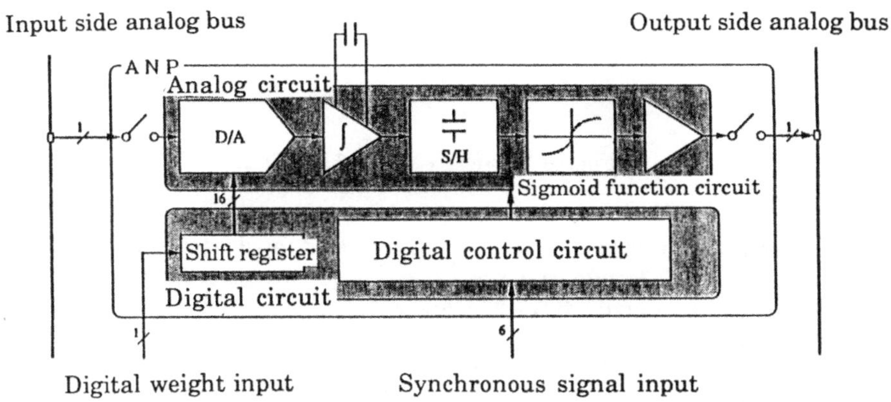

Figure 10. Analog neuron processor [5].

Applications

The field of neurocomputers promises to yield applications in control, pattern processing, and knowledge processing. Various applications are now being investigated.

Control This section explores the behavior control of a mobile robot as an example of an application of neurocomputers to the control field [1]. The mobile robots shown in Figure 11 are equipped with 12 sensors, and are controlled by neural networks. They must have flexibility such as learning and adaptation ability, and realtime processing functions for responding to

Figure 11. Mobile robot [1].

environmental changes. Conventional computers must perform the above functions by using sequential processing and must be programmed in advance so that signals from sensors can be processed in response to environmental changes. When the environment changes, conventional robots must ask the higher level controller to make a decision. In contrast, neurocomputers can respond flexibly to unexpected changes in the environment by using their learning and interpolating functions.

(1) Structured network model: To control the behavior of the robot by neural network, we adopted a structured network model which can handle a series of behavior patterns. Figure 12 shows the hierarchical network, which combines two types of layered networks, reason network and instinct network. The input signals from sensors and the output signals to motors are one-bit digital (0 or 1). Sensor signals are immediately fed to the input layer of the reason network. The reason network determines a correspondence between the sensory input and the behavior pattern, and outputs the behavior pattern. An example of such a behavior pattern is "move forward when the optical sensor at the head detects light." In contrast, the instinct network determines the correspondence between the internal state of the robot and the behavior pattern. An example is "when the robot is

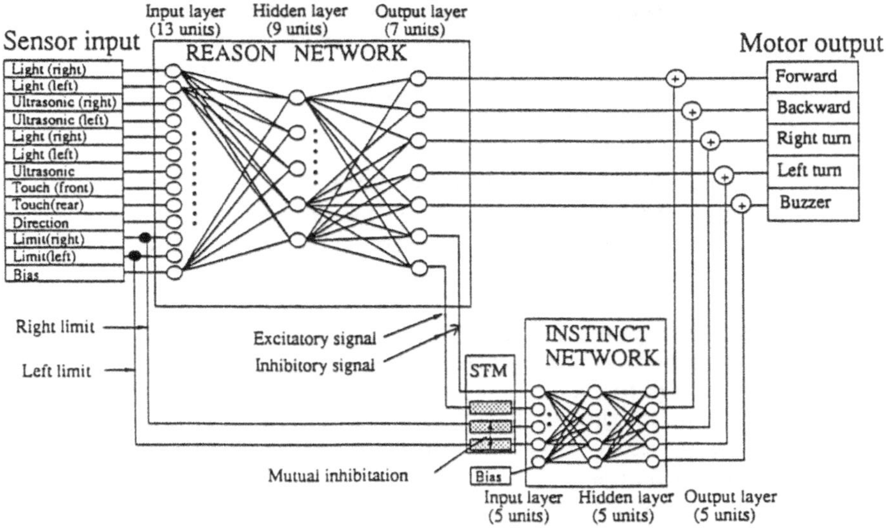

Figure 12. Neural network model used for mobile robot control [1].

excited after detecting obstacles, then move backward and turn left." As robots must often make such motions in sequence, the function of the instinct network is essential when applying a neural network to a robot.

(2) Short-term memories for motion in sequence: To process the internal state of the robot, short-term memory is connected to the input layer of the instinct network. A short-term memory unit is a unit which holds an excited or inhibited state for a certain period of time. The interaction between the reason and instinct networks is explained below. As illustrated in Figure 12, some of the output units of the reason network are interconnected to the instinct network by way of excitatory and inhibitory signals. The inhibitory connection from the reason network to the instinct network is reset only when there is no input signal from the sensors. The instinct network is taught to do nothing whenever inhibitory connection is active, it is why instinct network does not disturb the reason network's output.

When a signal from tactile sensor is received, the robot must take action to avoid the obstacle. This action is a behavior pattern series consisting of more than one motions in sequence - such as moving backward for a certain distance, then moving forward while avoiding the obstacle. For this action, the reason network transmits an excitatory signal to short-term memory unit, which holds the excitatory signal for a certain period of time. The instinct network continues to transmit a motor control signal for turning the robot and moving it in reverse as long as the short-term memory unit is excited.

When a limit sensor for robot torsion turns on, the instinct network is also activated. When the robot turns to the left or right and reaches the limit of rotation, a sensor signal corresponding to the limit position is transmitted to the pertinent input unit of the instinct network by way of short-term memory unit. The instinct network generates a motor control signal for right or left rotation in response to the left or right limit signal in order to restore the robot from torsion.

The robot behaves according to the reason network when it is given sensory information on which its action should be based, and behaves according to the instinct network when it is not given such sensor information, and must continue an action for a certain period of time.

It might seem possible that the above operation can be performed even though the network is not divided. However, without the hierarchical structure, the size of the network is so increased as to make realtime control difficult. In addition, the number of training patterns is so increased that raise the learning time. Division into two networks reflects the advance implementation of a known behavior in the neural network. The structured network helps reduce the number of training patterns and the number of the network connections, thus contributing to efficient learning.

(3) Multiphase learning: For more efficient learning of behavior patterns, it is necessary to consider the combination of teaching patterns so that the robot can respond in a reasonable manner when it is taught a minimum number of patterns. Our robot contains a relatively small number of units because a multi-hierarchical network is used; nevertheless, the number of pattern combinations that can be input to the reason network is 2^{12}=4096. If the robot must learn all these combinations, it is as cumbersome as when the robot is programmed and requires much time to teach the robot. The advantage of using a network is that only a minimum number of necessary patterns need be taught. The robot can determine its motion in any environment through inference from the patterns learned. We have adopted an efficient method called the multiphase learning method.

Learning is conducted in three phases. In the first phase, basic behavior patterns necessary for controlling the robot are created on a workstation using a network simulator and a behavior simulator which simulates sensor signals and robot motions, and the behavior is checked by simulation. In the second phase, detailed behavior patterns are created through tests on the actual robot connected to the workstation. In the third phase, network weights determined through learning are downloaded to the robot; the robot is then moved in stand-alone mode. If no problem is found, learning is complete.

This learning method raises the efficiency of learning because only the necessary patterns need be taught. Also, modification to patterns taught is so easy that robots with different behavior patterns can easily be created in a short period of time. In addition, the use of hierarchical networks allows both networks to be taught independently; this reduces the number of teaching patterns, contributing to a shorter learning time. In fact, we taught about 60 patterns to the instinct network; convergence was achieved with about 500 lessons.

(4) A scene of robot's tracking: A pursuit scene, shown in Figure 13, was played out by robots to verify the neural network's learning capability by demonstrating that robots can learn behavior patterns through a neural network according to input information from various sensors and can act appropriately in response to changing environment conditions. Another purpose is to verify the flexibility of the neural network by building robots with different behavior patterns by changing the correspondences between sensor input information and behavior patterns from a single robot.

First, we fabricated robots with two different types of behavior pattern; "capture" and "escape" - that is, robots with a behavior pattern for tracking other robots emitting infrared rays or ultrasonic waves and robots which escape when they sense these rays. These robots can be made by teaching different behavior patterns to robots with the same physical structure. Capture robots search for ultrasonic waves and infrared rays from escape robots and track the escape robots. Escape robots escape from rays being emitted by search robots. Each robot proved to be capable of determining its motion in response to other robot's motion according to the behavior patterns it learned or using these patterns to infer the correct action if input information does not match any patterns exactly. Although the robots were not equipped for direct communications, they acted as if they were cooperating and coordinating their motion. The scene was like a behavior simulation of animals. The robot's behavior seems to be based on a representation internal to the hidden-layer unit. The development of this system has demonstrated the usefulness of a structured network.

Figure 13. Mobile robots controlled by neural network [1].

Pattern processing The application of neurocomputers to the field of pattern recognition in such areas as character recognition and speech recognition is widely being advanced. Recognition can be easily be implemented via learning, and recognition results that match the intuition of a human being can be obtained. The applications in the area of computer graphics is also being investigated.

Knowledge processing The applications of neurocomputers to this field are mostly concerned with pattern recognition, and can be expected to yield the same advantages as in the preceding section. Figure 14 shows an example of an application of knowledge processing to bond ratings. This example obtains relatively good judgement results.

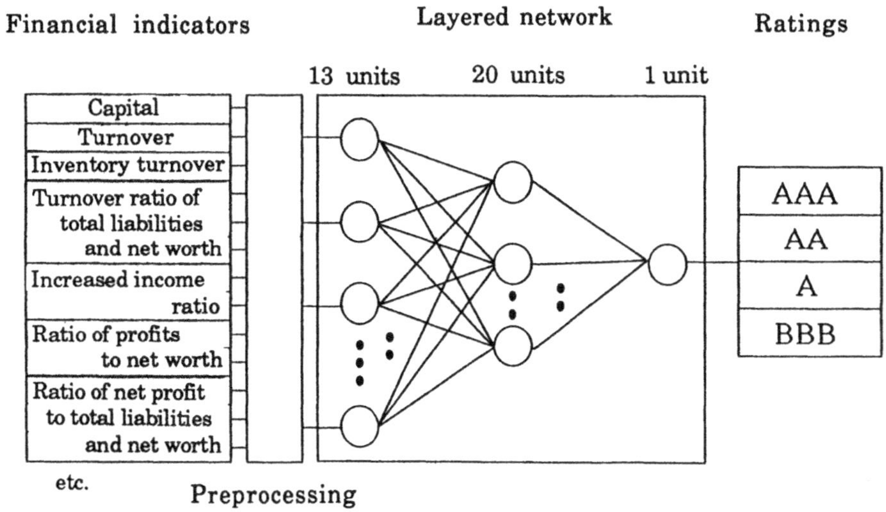

Figure 14. Example of application to bond ratings.

CURRENT STATE OF RESEARCH ON NEUROCOMPUTERS IN JAPAN

In Japan, research and development of neurocomputers is being conducted by many large companies in the neural network models and their learning algorithms, implementation tools, and applications. Table 1 lists the typical research trends on neurocomputers in Japan.

Amari's contribution to Theory

The boom in Perceptron research had passed and research on neurocomputers hit a low in 1970. In Japan, research on the basic principles, possibilities, and limits of neural networks has steadily been continued by Amari and several other researchers. This is in marked contrast to the neurocomputer research toward direct application and information processing systems.

Table 1. Current research activities in Japan.

Manufactures	Research activities	Remarks
NEC	Speech recognition	DNN
	Character recognition	Multi-fonts
	Dedicated parallel simulator	MANYO, NeuMan
	Optical computer device	1k byte optical device VSTEP
Hitachi	Neuro-portofolio	Optimization problem
	Speech recognition	Coupling to fuzzy system
	Robot control	Reinforcement learning
Mitsubishi Electric	Character recognition	Optical associative computer
	Optical neuro chip	Compactness
	Robot control	Crebellum by Albus
Toshiba	Task allocation for parallel computer	Optimization problem
ATR	Perception	Modeling of organism functions
	Robot control	
	Large-scale neural network simulator	NCUBE
	Speech recognition	TDNN
	Noise elimination from speech data	
NTT	Medical diagnostic expert system	Aquisition of rules
Other Manufacturers	Matsushita Electric, Canon, Nippon Steel Corp.,Asahi Kasei, NHK	
Universities	University of Tokyo, Toyohashi University of Technology	
Institutes	Electrotechnical Laboratory	

Notably, Amari has shown the specific solutions to nerve fields with simple structures, and has also considered the case in which two or more nerve fields are coupled one after another. He has also shown that information processing by competition and cooperation of nerve cells is made possible by combining such uniform nerve fields, as illustrated by the proposed models of the worm detection network in frogs, and the binocular stereoscopic vision network [6]. In addition, he formulated the nerve fields in which the nerve cells are provided with the learning function, and also mathematically analyzed the mechanism by which the topography of information structure of the external world is reflected on the insides of neural networks [7]. This is assumed to explain the topography of a nerve system such as retinotopic mapping. Among his many accomplishments that make use of mathematical methods are his research on the self-formation of feature extraction cells, his research on the learning method of associative memory, and his statistical dynamics theory on the behavior of nerve fields that are mutually coupled at random. It is also extremely interesting that he proposed during the time of Perceptron research in the 1960s the same principle as that used by error back propagation method whose effectiveness is now demonstrated in various applications [8].

Amari's basic research will play an important part in advancing neurocomputers in the 1990s.

Application

Trajectory control of a robotic manipulator This section introduces the control of manipulators as an application of neurocomputers to robot control. To smoothly control the movement of a complex mechanism such as a manipulator, the model-based control, that incorporates a mechanical model of manipulator into a computer and calculates control commands based on it, becomes important, in addition to feedback control. This control method requires a processor that can execute complex operations at high speed. The model-based control method, developed as a manipulator-

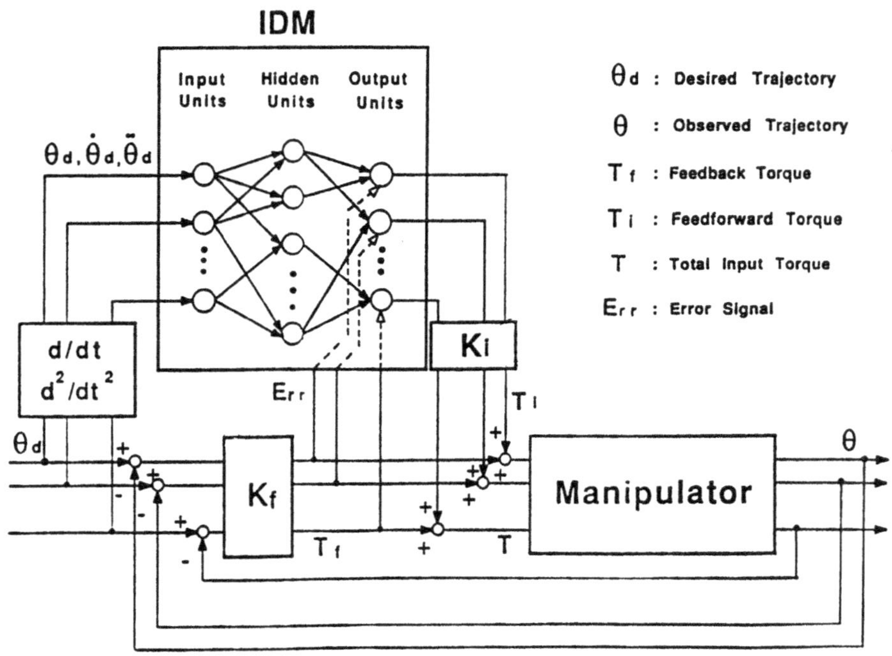

Figure 15: Acquisition of inverse model by using a neural network [9].

specific control method, was not emphasized in the formulation of conventional control theory, in which the feedback control system is mainly used.

The dynamics model is used to improve the dynamic characteristics of control systems. Inverse dynamics is used to obtain the torque that controls the manipulator in a specified movement trajectory. The torque is obtained by substituting the position, velocity, and acceleration calculated from the trajectory into the dynamics equation for the manipulator. The dynamics equation of the manipulator is generally expressed in the following equation.

$$M(q)\ddot{q} + h(q, \dot{q}) + \Gamma\dot{q} + g(q) = \tau.$$

Where, $q \in R^n$ is a joint angle vector that express the displacement of the joint angles in n-dimensional Euclidean space, $\tau \in R$ is a driving force vector, $M(q)$ is an inertia matrix, $h(q, \dot{q})$ is centripetal and Coriolis force, Γ is a matrix of friction force, and $g(q)$ is a gravitational force. The calculation of inverse dynamics is generally complicated, and its realtime calculation especially difficult.

Figure 15 shows an example of acquiring the inverse-dynamics model by neural networks [9]. In addition, a neural network that generates the trajectory of manipulator with smallest torque is being studied [10].

<u>Optical neurocomputer</u> An attempt is being made to design neurocomputers that use optical technology. If optical technology is used, the basic operations of neural networks such as multiplication and summation are easily

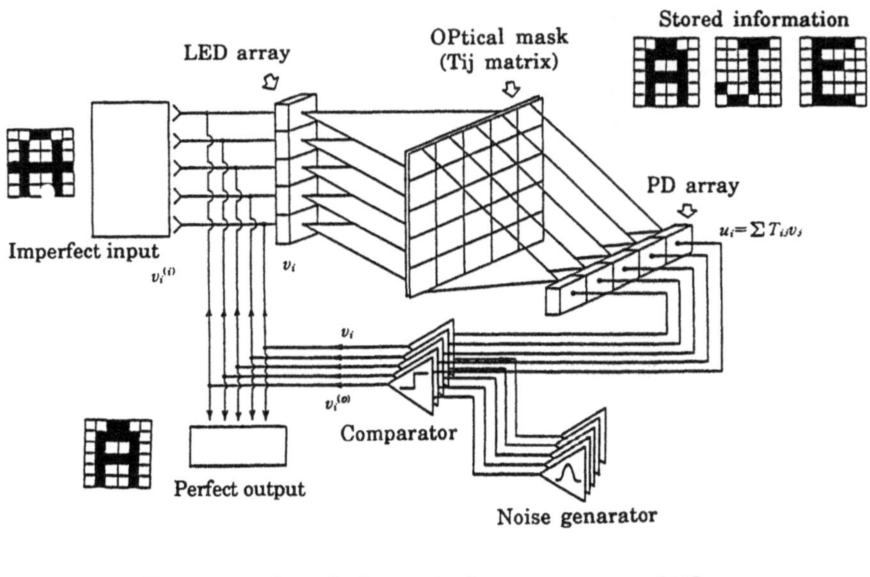

Figure 16. Associative optical neurocomputer [11].

implemented. The complicated interconnections for VLSI can also be simplified.

Figure 16 shows an optical associative memory system into which the idea of simulated annealing has been introduced [11]. This system, containing N(=32) neurons and M(=3) memory vectors, uses an optical matrix-vector multiplier as the basic structure. The activation level of each neuron corresponds to the state of one LED. The signals, photoelectrically converted by photodiode arrays, are processed for the threshold values by comparators and are fed back to the LED arrays. A, J, or E is selected for each imperfect input. Perfect output can be obtained with repeated operation. Matrix Tij is bipolar with positive and negative components that correspond to excitatory or inhibitory synapses.

CONCLUSION

This paper outlines the current status of research on neurocomputers in Japan. Emphasis is placed on research conducted by Fujitsu.

To advance neurocomputer research, emphasis must be placed on basic research, that pursues the basic principles, possibilities, and limitations of neural networks, in addition to specific neurocomputer applications research.

The neurocomputer may be said to have a complimentary relationship with conventional computer. The conventional computer is characterized by its ability to perform the functions associated with the left brain: logic processing and numeric calculation. Neurocomputer, on the other hand, are characterized by their ability perform functions of the right brain: pattern processing and intuition. Extremely flexible, intelligent computers that are nearer to human beings will be possible when the advantages of the conventional computer and those of neurocomputer are fully merged.

REFERENCES

1. Watanabe, N., Nagata S., and Asakawa, K. Mobile Robot Control by Neural Networks and Their Associated Learning Algorithm. Proc. IEEE/IECON'88, pp.1054-1060, 1988.

2. McClelland, J.L., Rumelhart, D.E., and the PDP research group. Parallel Distributed Processing, Vol.1 and Vol.2, MIT Press, Cambridge, 1986.

3. Nagata, S., Kimoto, T., and Asakawa, K. Control of Mobile Robots with Neural Networks, Neural Networks, Vol.1, Sup.1, p.349, 1988.

4. Masuoka, R. A Study on Fast Learning Algorithm of Back Propagation, (in Japanese), IEICE MBE 88-176, 1989.

5. Hijiya, T., Sugiura, Y., Iwamoto, H., Yoshizawa, H., Kato, H., and Asakawa, K. Nikkei Microdevices, (in Japanese), No.45, pp.123-129, 1989.

6. Amari, S. and Arbib, M.A. Competition and Cooperation in Neural Nets, Systems Neuroscience, (Ed. Metzler, J.), pp.119-165, Academic Press, New York, 1977.

7. Takeuchi, A. and Amari, S. Formation of Topographic Maps and Columnar Microstructures in Nerve Fields, Biological Cybernetics, Vol.35, pp.63-72, 1979.

8. Amari, S. A Theory of Adaptive Pattern Classifiers, IEEE Transactions on Electronic Computers, Vol.16, No.3, pp.299-307, 1969.

9. Setoyama, S., Kawato, M., and Suzuki, R. Manipulator Control by Inverse-dynamics Model Learned in Multi-Layer Neural Network, (in Japanese), IEICE MBE87-135, 1988.

10. Kawato, M., Uno, Y., Isobe, M., and Suzuki, R. A Hierarchical Neural Network Model for Voluntary Movement with Application to Robotics, IEEE Control Systems Magazine, Vol.8, No.2, pp.8-16, 1988.

11. Ohta, J., Tai, K., Kuroda, K., Oita, M., Kyuma, K., and Hamanaka, K. Associative Optical Computer, (in Japanese), IEICE OQE87-174, 1988.

Advanced Communication Technology and OSI Oriented Academic Network TAINS

S. Noguchi

Computer Center, Tohoku University,

2-1-1 Katahira, Sendai, Japan

ABSTRACT

Recognizing the profound impact of communication technology on the progress of human civilisation, a major thrust in research and development in this area is being made all over the globe. In this paper, glimpses of the rapid strides that are are being made in this direction, in Japan, are presented. In the first part the various technological developments, such as VLSI, optical devices & communication, computer systems, ISDN etc., which have a direct or indirect bearing on communication technology is reported. In the second part a pioneering step in the direction of a futuristic information exchange network - an OSI oriented large scale academic network- TAINS, which utilizes the advanced communication technology and concepts , is described.

INTRODUCTION

In the coming information age communication technology will play a vital role in accelerating the progress of human civilisation. Recognizing this profound impact of communication technology, a major thrust in research and development in this area is being made all over the globe. Japan is playing a significant role in the global search for the technology that will answer the needs of the present and the future.

In this paper, glimpses of the rapid strides that are are being made in this direction, in Japan, are presented. In the first part the various technological developments, such as VLSI, optical devices & communication, computer systems, ISDN etc., which have a direct or indirect bearing on the growth of communication technology is reported. In the second part a pioneering step in the direction of a futuristic information exchange network - an OSI oriented large scale academic network- TAINS, now operational in Tohoku University, which utilizes the advanced communication technology and concepts , is described.

The areas of modern communication research can be abstractly represented in the hierarchical diagram shown in fig. 1.

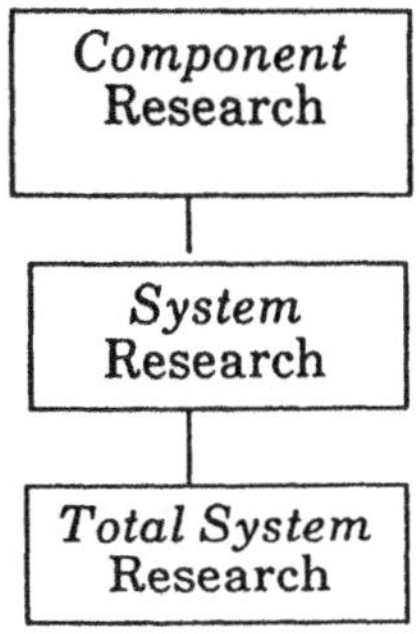

Fig 1

A *total system*, for example ISDN, can be designed by employing the advanced *system* design technology such as that of large scale switching systems, computer systems, etc.. On the other hand an advanced *system* can be designed by utilizing the *component* technology such as that of VLSI's and optical devices & communication. Typical items in each layer are given in the following.

(1) *Component* Research
(a) Optical devices & communication
(b) VLSI's

(2) *System* Research
(a) Intelligent interfaces
(b) Local area Networks
(c) Large scale switching systems
(d) Large scale computers

(3) *Total System* Research
ISDN

In this report, representative items are selected from each of the above layers in the hierarchy. Nevertheless, it must be mentioned that the samples cited span a very limited area and this report does not in any way amount to a comprehensive survey of the present state of communications research in Japan.

I Advanced Communication Technology

1. Optical devices & communication.

(1) Optical fiber transmission[1] Scientists at the research laboratories of NEC corporation have reported the successful execution of a 10Gb/s - 100Km optical fiber transmission experiment using a high speed MSQDFB-LD and back illuminated InGaAs-APD. The experimental set-up is shown in fig 2. It uses a MQW DFB-LDÅ as transmitter and a back illuminated InGaAs-APD as receiver. They have reportedly solved the problems of insufficient device frequency response, chirping, low receiver sensitivity, etc.

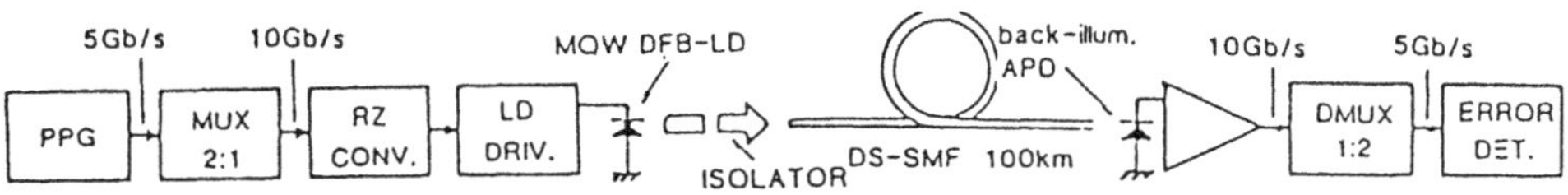

Fig 2. Block diagram of the 10 Gb/s transmission experimental setup

The LD utilized in the experiment is a 1.5 μm λ/4 shifted MQW DFB-LD with 10 InGaAs wells (75 Å thick) and InGaAsP barriers (150 Å thick). The relaxation oscillation frequency of the LD is 1.5 times higher than that of a regular bulk DFB-LD as is shown in fig 3. This result indicates the feasibility of 10Gb/s transmission.

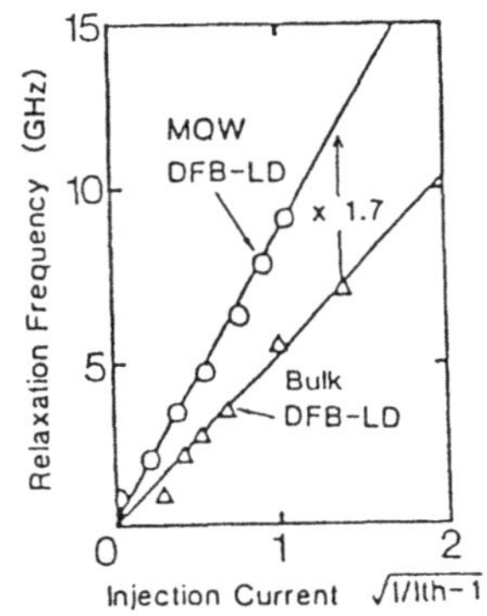

DFB-LD resonant frequency dependence on injection level

Fig 3.

The total dispersion for the DS fiber used was 37ps/nm at the DFB wave length. The measured gain-bandwidth product for the back illuminated APD was 60GHz and a receiver sensitivity of 23.5 dBm was attained. The bit error rate characteristics for the 10Gb/s - 100.1 Km transmission is given in fig 4.

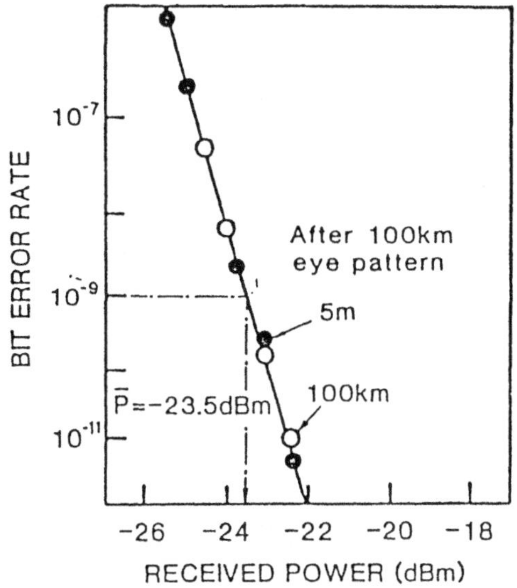

Fig 4.

(2) Photonic switching[2] The Research group of NEC Corporation has announced a photonic space division switching system using 8×8 polarizaton independent $LiNbO_3$ matrix switches, for High Definition TV services. This experiment explores the feasibility of large scale photonic switching systems using ordinary single-mode optical fibers.

The system configuration of this experiment is shown in fig 5. The configuration of the 8×8 matrix switch is shown in fig 6. In this experiment, the optical loss of the external splitter using fiber couplers and that of the third stage switch were 10dB and 13-15dB, respectively. As for the TV phone services, the total insertion loss of the 3-stage switching network was 28~33 dB. In this experiment, the HDTV analog signal is encoded to 32.4 Mb/s, 19 parallel streams by an encoder and converted to 780 Mb/s serial service signal. As the conclusion, it is reported that 5 ~ 10 dB system margins with 2 Km subscriber lines are expected for TV phone services.

This experiment is another step forward in the direction of practical realization of the photonic space division switching system using $LiNbO_3$ matrix switches.

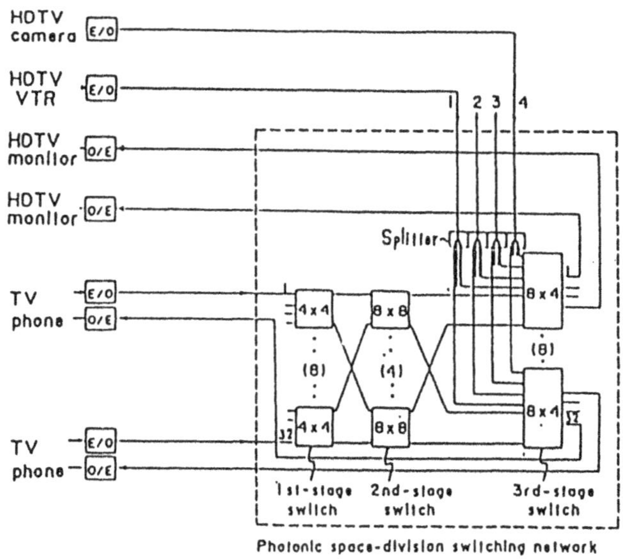

Fig 5. Photonic switching system block diagram

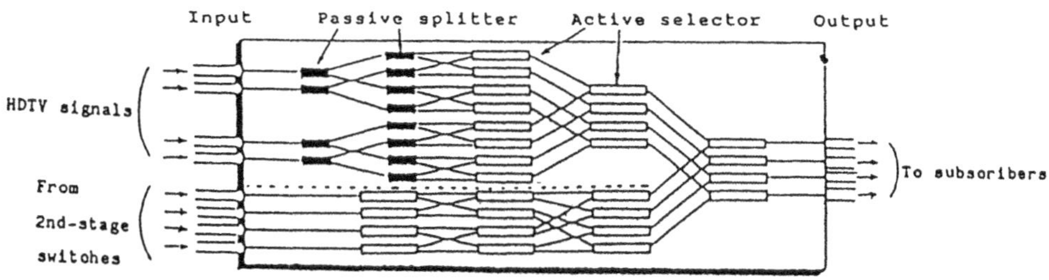

Polarization independent 8X4 matrix switch for 3rd-stage switches

Fig 6.

2 VLSI Technology

In the field of VLSI technology, Japan has been in the fore-front. From among the several remarkable results, two important products are reported in the following.

(1) SPARC CHIP[3] The Matsushita research group has announced the development of the SPARC chip MN10501 this year. This chip is fabricated on a 15×15 mm^2 substrate utilising CMOS technology of 0.8 µm rule. This chip contains a 64-bit floating-point arithmetic unit and has the processing speed of 25MIPS (VAX-11/780 base). The heat dissipation is 4W.

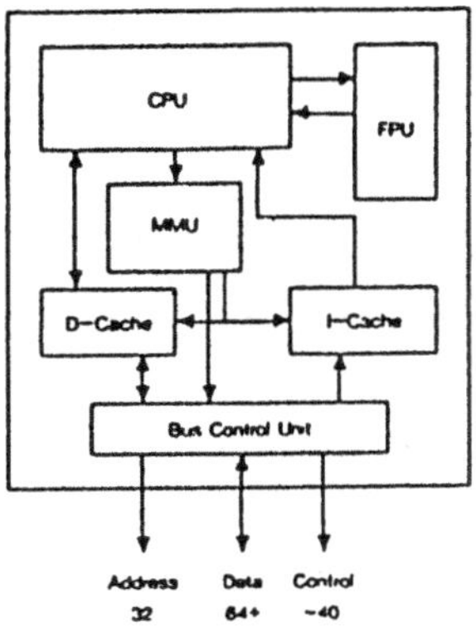

Fig 7.

In order to increase the data transmission capability some part of the inner and outer bus of the chip has a 64 bit data bus but the CPU core has a 32-bit architecture in order to make it compatible with existing SPARC chips. The block diagram of the chip is shown in fig 7.

(2) HEMT[4] High Electron Mobility Transistors (HEMT) are very promising devices for ultra high speed and low heat-dissipation VLSI's. Fujitsu has been developing this transistor since 1980. At the present stage, the switching delay of HEMT's is lower than 10ps which makes it a very likely candidate for devices which will be used in the next generation of super computers.

A 4.1K-gate HEMT array with a 16×16 bit parallel multiplier has been developed to achieve a multiplication cycle of 4.1ns. A 4K-bit HEMT static RAM with access time of 500ps and a 16K-bit HEMT static RAM with access time of 3.4ns have been developed. An optimised chip delay of

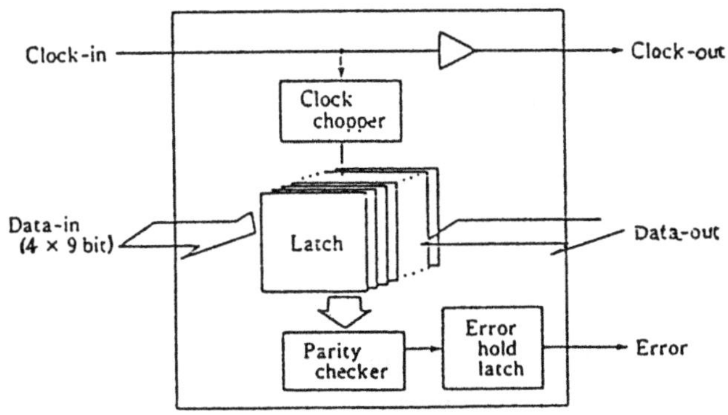

Fig 8. Block diagram of a multibit data register circuit.

70ps for a 10K-gate HEMT VLSI at room temperature is being expected. A multi-bit HEMT data register will find place in supercomputers in the near future. A block diagram of the multibit data register and its performance are shown in fig 8 and table 1.

Minimum length	0.5 microns
Gate counts	1.1K gates
Chip size	6.1 mm × 6.2 mm
Basic gate delay	19 ps
Clock-in to Data-out	490 ps
Power disspation	4.1W

Table 1

3 Intelligent interface[5]

Several intelligent interfaces are being developed to suit applications. One interesting intelligent interface is a mechanical translator. This is the worlds first computerized Japanese to English speech-interpreter system developed by Matsushita research group. This system can recognize and understand the Japanese speech of any speaker. It can perform rapid interpretation. For example, a two clause Japanese conversation is interpreted into English in ~ 5 seconds.

The main function of speech processing can be carried out on any 32bit workstation. This system adopts a newly developed technology that recognizes speech and converts it into phoneme symbols. Presently, the system is utilized in a restricted environment but more general usage is being targeted. The system will be of portable size. An outline of the flow of information processing is given in fig 9.

4 Local Area Network (LAN)[6][7]

With the rapid increase of information utilization, the demands on LAN has become very complex. LAN must support very high speed data transmission in conjunction with several kinds of multi-media transmission. NEC corporation and Fujitsu Ltd have developed very large capacity multi-media LAN which have an operating bit-rate of 400Mbps.

These are the NEC C&C NET LOOP6560 and the Fujitsu FACOM 2890. These LAN's can integrate low speed data, high speed data, voice data and video data and, are used as the backbone LAN which supports several LAN's for network integration. Using advanced optical technologies LOOP 6560 and FACOM 2890 provide 400Mbps and 410Mbps transmission, respectively .

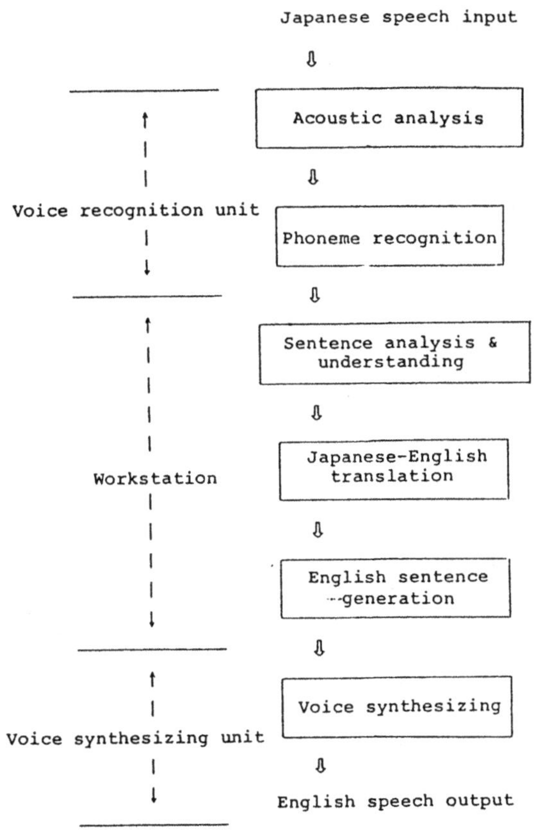

Fig 9.

The key technologies which have led to the success of the large capacity multi-media transmission are as folows:

1. High speed optical transmission
2. Mini-packet (ATM) processing for multimedia communication
3. Very high speed VLSI communication processors for message exchange.
4. Distributed processing software.

A typical LAN configuration for multi-media data highway is shown in fig 10. Fig 11 shows the mechanism of high speed parallel packet processing by FACOM 2890.

The specification of both LAN systems has almost the same characteristics. The specification of the channel capacity of LOOP 6550 is shown Table 2.

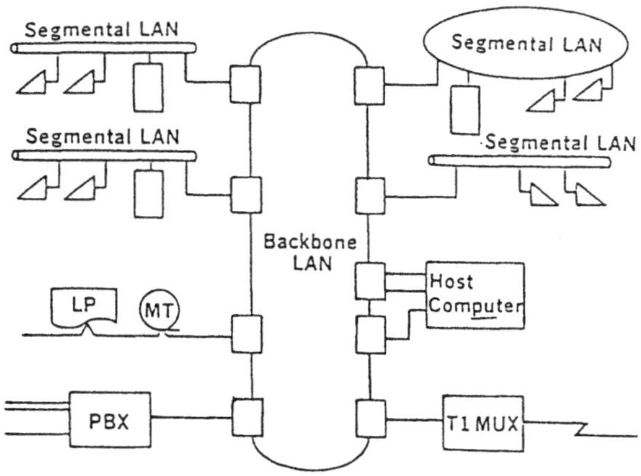

Fig 10.

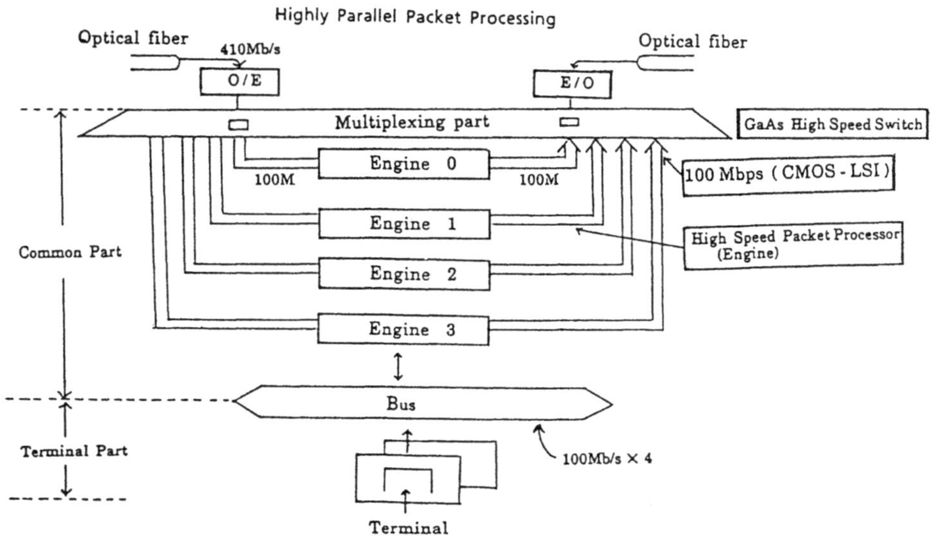

Fig 11.

all interfaces are CCITT, TTC, NTSC standards. Mixed mode utilization of the channels is available at a channel capacity of 400Mbps.

The total number of nodes contained in this network is 64/loop and maximum distance between nodes is 10 kms.

5 ISDN[8]

Currently NTT is the largest common carrier in the country and plays the dominant role in the promotion of communication technology in Japan. The ISDN project promoted by NTT is reported in the following.

INTERFACE	NUMBER of CHANNELS
100M (MP*)	4
45M(MP)	4
32M(MP)	8
Bridge	28
1.536M	248
768	496
384	988
192	1,964
64K(Speed)	5,712

* MP (Moving Picture)

TABLE 2.Channel capacity of LOOP 6550

(1) Present Network services

The existing tele-communication network and services offered by NTT are shown in fig. 12. Though several new comon carriers are expanding

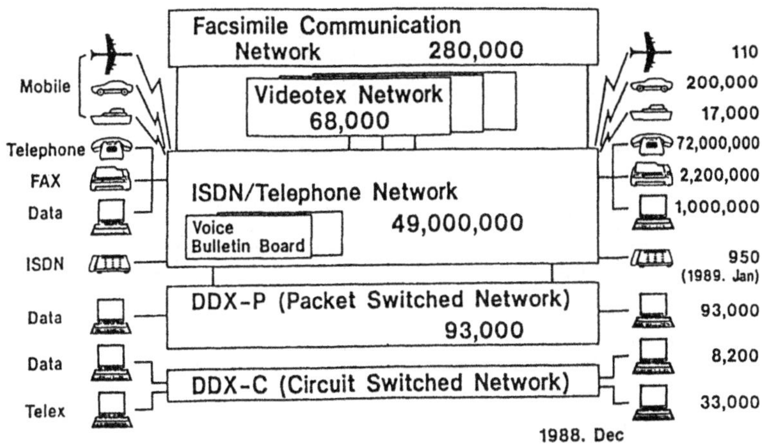

Fig 12.

their services very rapidly, the NTT network and its services are still dominant. In April 1968, NTT started the ISDN services called INS64. This is the basic ISDN and only supports a 64Kb/s service. But, this year NTT is launching the INS 1500 which will support 1.5Mb/s services. Against this backdrop NTT has inaugurated the project of the information network for the 21st century.

The two important keywords in this project are (a) Universal connectivity and (b) Intelligent services in a Multi-media environment.

The former emphasizes the importance of participation in and cooperation with international activity while the latter focusses on the adoption of new technology developed in the fields of logic, AI and cognition science etc.. The central ideas in the new information network that is the future ISDN are outlined in the following.

(2) Concept of future ISDN architecture

New architectures are designed according to the following concepts-

(a)Define the total network system in a logical scheme and describe its function using formal specifications

(b) Decompose the total system into several hierarchical subsystems according to the functions of the network service.

(c) Represent the function of each layer by a basic module.

(d) Define clearly the interface between layers and the protocol between modules.

The ISDN structured architecture is shown in fig. 13.

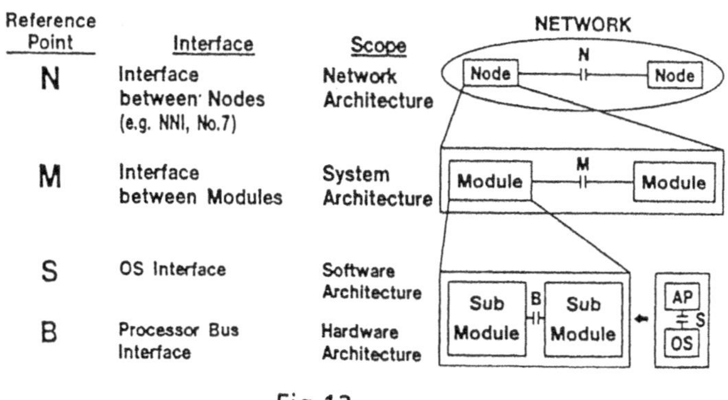

Fig 13.

This diagram consists of the following four layers:

- Network layer
- System layer
- Software layer
- Hardware layer

The interface reference point between layers and modules is also shown in this figure.

The diagram given here is the first level representation of the architecture scheme. At the next step, it is necessary to decompose it into a finer logical structure according to the above principle.

The specification of the network is focussed on and is explained in the following.

(3) Network Layer Architecture
The Network layer has two functions these are (a) support of the fundamental information transportation such as switching and transmission and (b) support of enhanced service such as network management and service control. Accordingly the network architecture basic model shown in fig 14 is derived.

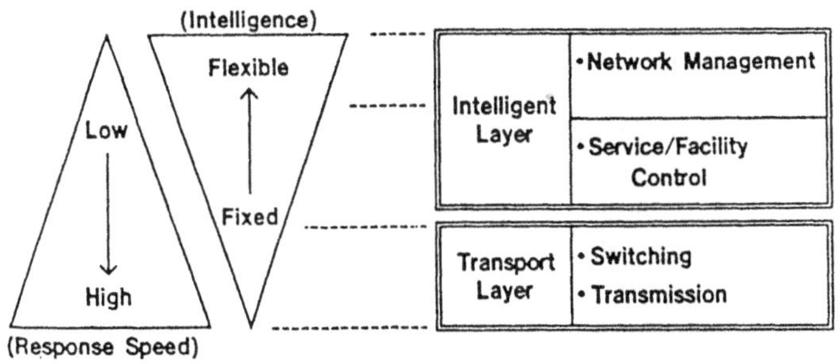

Fig 14.

The model comprises of two layers called called the transport layer and intelligent layer, respectively.

(4) Transport Layer Architecture
The transport layer supports the fundamental user information transmission and it must support the three key functions. These are (a) control of physical transmission on media (b) support of the bundles of standard circuit and (c) the circuit service of user directly.

Accordingly, the transport layer is decomposed into the folowing three layers as is shown in fig 15.

(a) Circuit network layer
(b) Path network layer
(c) Transmission media network layer

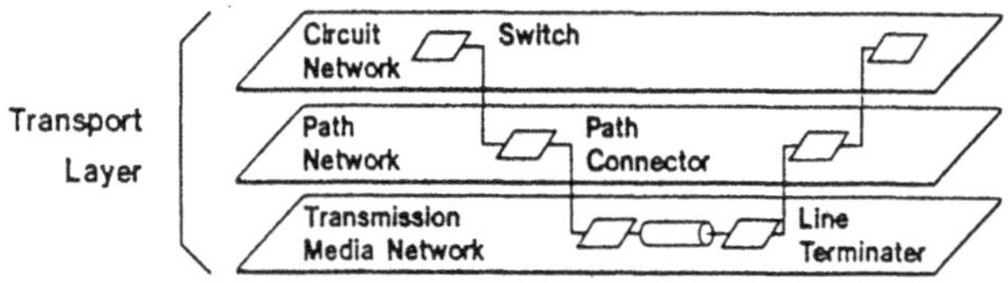

Fig 15.

The key functions of each layer is as follows :

(a) Transmission media layer: support of the operation, administration and management (OAM) of the physical media such as optical fiber μ-wave and coaxial cable etc.

(b) Path layer : support of the standard bundle of circuits in ISDN such as 1.5 Mb/s , 52Mb/s and so on by utilizing the services of the transmission media layer. The corresponding OAM activities are also supported in the layer.

(c) Circuit layer: supports the service of users requirements in ISDN such as 64K/s , 1.5Mb/s services and so on by utilizing the services of the path layer. The corresponding OAM activities are also supported in the layer.

System architecture, software architecture and hardware architecture shown in fig 12 are also decomposed intc subsystem architectures in a similar manner. The C-TRON project now being promoted in Japan concerns the OS level architecture. But, these details are omitted here.

II OSI Oriented Academic Network TAINS

Tohoku University has been constructing the academic information-exchange network TAINS (Tohoku University Academic/All-round

/Advanced Information Network system) to link its widely spread campuses to meet the demands of an intra-university network. TAINS, the first network based on OSI became operational in April, 1988.

The primary function of TAINS is to connect all the geographically dispersed locations of the university via a high speed digital network. In order to facilitate an easy and efficient interconnection among the various computers, terminals and other scientific equipment, a trunk network and several inhouse networks are constructed. Each inhouse network employs the 10Mbps IEEE 802.3 protocol and inter-connects the computers and equipment in a laboratory or office.

1 Design concept of TAINS

The general goals pursued in the design of TAINS are

(1) a network of highest quality using state of the art communication technology,

(2) a network that can be upgraded in step with the developments in science technology.

(3) a network that will, in future, enable the exchange of information among universities and research institutions in the country and overseas too.

2 Expectation from TAINS

The requirements for TAINS to support academic activities were investigated and were categorized as follows.

(1) database retreival
(2) on-line execution of scientific computations
(3) document processing
(4) processing of experimental data
(5) image data processing
(6) electronic mail exchange
(7) communication among personal computers
(8) video-tex
(9) TV-lectures

(10) TV-conference
(11) library information
(12) connection with networks outside the university
(13) execution of CAD/CAM/CAE

Based on these requirements, the configuration of the network was considered. But due to technical and economical constraints, the functions related to image telecasting ((9), (10)) were shelved to be realized at a future date.Moreover, from the projected increase in high-speed and large scale data transfer in activities related to image data processing, scientific data processing and large scale scientific calculation, TAINS had to aim at high-speed data transmission.

3 Construction of the communication network in TAINS

A two layer network configuration was adopted for TAINS as is shown in Table 3.1 and Table 3.2. The lower layer consists of the inhouse networks which use coaxial cables and generally span a building. The second layer consists of an optical fiber trunk loop which connects the inhouse networks. Basic functions of the TAINS inhouse networks and the trunk network are listed in Table 3.1 and Table 3.2, respectively. The inhouse network is a bus type network based on IEEE802.3(CSMA/CD) and is set in the building of each faculty, research institute and so on. The trunk network is constructed by connecting the node control unit of each inhouse network. It is a 100Mbps, token-passing ring network and uses multi-mode optical fiber. The network has built-in network diagnosis, backup, self loopback, and other features. The total length of the optical fiber for the trunk network is ~ 35 Kms and the length of the coaxial cables used in the inhouse networks is ~40 Kms.

4 Protocol in TAINS

(1) Inhouse network The inhouse network protocol is listed in Table 4. The physical and MAC layer protocol is based on 10Mbps, IEEE802/3 protocol and, coaxial cable is used. In the third layer i.e. the network layer the connectionless mode, ISO IS8472 protocol is adopted, so that that all TAINS users are able to access any terminals from any terminal at all times and from all places. In the LLC layer, ISO IS802/2 type1

Netwok Configuration	Branch Network
Media	Coaxial Cable Optical Fiber(REMOTE REPEATER)
Transmission Speed	10Mbps
Transmission Method	Baseband
Number of Stations	max100/segmentT max1025/network
Station Separation	max2500m, min2.5m
Cable Length	max500m/segment
Media Access Method	CSMA/CD(ISO 8802/3)

Table3.1 Basic Function of Inhouse Network

Netwok Configuration	Loop
Media	Optical Fiber
Transmission Speed	100Mbps
Transmission Method	Baseband
Max Number of Nodes	126
Standard Node Distance	2km
Media Access Method	Token Passing Ring FDDI of ANSI
Interface	ISO8802/3
Measure to counter trouble	Self-diagnosis, Healthcheck, Bypass, Loop-back and Alternative Functions

Table3.2 Basic Functions of Trunk Network

Application Layer	based upon OSI(MHS,FTAM,etc)
Presentation Layer	based upon OSI
Session Layer	ISO IS8327
Transport Layer	ISO IS8073 Class4
Network Layer	ISO IS8473 (Connectionless mode)
Data Link Layer	
LLC	ISO IS8802/2 Type1 (IEEE802.2)
MAC	ISO IS8802/3 (IEEE802.3) CSMA/CD
Physical Layer	10BASE5, Coaxial Cable(50Ω)

Table4 TAINS Inhouse network protocol

protocol is adopted, as this matches with the connectionless mode protocol of the network layer. The transport layer protocol defines the communication quality. In TAINS the highest level protocol IS8073 class 4 is adopted, The fifth layer i.e. the session layer, will be fixed considering

the upper layer protocol. At present, only the basic functions are selected. Standards for the presentation layer and application layer protocols are now in the process of being set by ISO, once the standards are announced, the protocols for the last two layers will be fixed for TAINS.

(2) Trunk network The trunk network is constructed by connecting the node control equipment set for each inhouse network with the multi-mode optical fiber. The trunk network is a 100Mbps, token passing network which has adopted the FDDI (Fiber Distributed Data Interface) of ANSI. The node control equipment is designed to carry out the addressing using only MAC addresses.As TAINS has the appearance of one large IEEE802.3 network and addressing is carried out using MAC addresses, though it is based on the OSI protocol, other network architectures such as TCP/IP XNS, DECNET, DINA also can co-exist in the network.

(3) Connection of low speed terminal to TAINS: development of the Communication server It is quite important for the development of TAINS to allow easy connection of personal computers to TAINS and to make possible the easy inter-connection between personal computers and host computers. The lower layer protocol of TAINS is basically IEEE802.3, but at present there are few personal computers that have this IEEE802.3 interface. So the connection of the personal computers to TAINS needs something like a communication adaptor. In TAINS, a new communication adaptor named CS (Communication Server) has been developed. The interface conditions of CS are listed in Table 5. A

Interface Specification		
	Pysical Interface	CCITT V.24/RS-232C
Terminal Side	Protocol	Nonprotocol HDLC/SDLC
	Number of Ports	10
	Line Speed	300~19,200bps
	Interface	ISO 8802/3(IEEE802.3)
Inhouse Network Side	Protocol	OSI
	Transmission Speed	10Mbps

Table 5 Interface condition of Communication Server

maximum of 10 personal computers may be connected to a CS and the maximum transmission speed between a personal computer and the CS is 19.2Kbps. CS follows the standard TAINS protocol, and as the virtual terminal protocol TELNET is followed. Currently, TAINS supports almost 5,000 terminals, in the future this figure is expected to rise to 10,000 terminals.

4 Protocol matching problem TAINS is based on the OSI protocol. But at present various kinds of network architectures exist in Tohoku University. The inter-connection between these different architectures

has to be be solved. In this context work is in progress in the following directions.

(a) Connection between OSI system and UNIX system.

(b) Connection with networks outside the university.

5 Present and Future of TAINS.

Although not all the functions defined in the section II.2 are available at present, all functions are expected to be installed in the near future. The present services are as follows.

(1) Utilization of personal computers as terminals.

(a)Utilization of host comnputers

In Tohoku University the following host computers are existing.

- General purpose main-frame computer ACOS 2020 (90 MIPS at the Computer center)
- Supercomputer SX-2N (1.3 GFLOPS, Computer center)
- Educational purpose large computer IBM3081 (16 MIPS, at the Education center for information processing)
- Several medium size computers.

(b) Utilization of UNIX machines

(c) Communication among personal computers

(2) Communication among workstations: TAINS adopts the ETHERNET-TCP/IP protocol as the quasi-standard protocol, so workstations based on this network architecture can use the remote login and FTP functions through TAINS. The transmission rate upper bound is around 1Mbps.

(3) Utilization of the Image data transfer system: The image data transfer system consists of an image data transmitter and a receiver installed at each inhouse network site and, an image mail server. By using this system documents of 400DPI quality can be easily exchanged via TAINS. TAINS is designed based on a new concept and is the first large scale campus network in Japan. It is expected that TAINS will continue grow and contribute to the development of research and eduation in Tohoku University.

Epilogue

Glimpses of the rapid strides that are are being made in research and development in the area of information communication is presented. In the first part the various technological developments, such as VLSI, optical devices, computers systems, ISDN etc., which have a direct or indirect bearing on the growth of communication technology is reported. In the second part a pioneering step in the direction of a futuristic information exchange network - an OSI oriented large scale academic network- TAINS, which utilizes the advanced communication technology and concepts, is described. It is well recognized that developments in communication technolgy will have a far reaching effect in the development and well being of peoples all over the globe and, in promoting peace, prosperity and mutual understanding among the peoples of the world. The need of the hour is to promote research activities in this area in an atmosphere of friendship and cooperation.

Acknowledgement The author would like to thank the various research groups in the communication and computer industry, specially to those in NTT, NEC, Hitachi, Fujitsu and Matsushita. Without their hearty cooperation this paper would have been less informative and less interesting. The author would also like to extend his thanks to Dr. Glenn Mansfield who has helped in making a better presentation of the material in this paper.

REFERENCES

[1] S. Fujita et.al. *A 10Gb/s -100Km Optical Fiber Transmission Experiment Using a High Speed MQW DFB-LD and A Back Illuminated InGaAs-APD* in Topical meeting on Picosecond Electronics and Optoelectronics, Salt Lake City, March 1989.

[2] S.Suzuki et.al. *HDTV Photonic Space Division Switching System Using 8 × 8 Polarization Independent $LiNbO_3$ Matrix Switches* in Topical meeting on Picosecond Electronics and Optoelectronics, Salt Lake City, March 1989.

[3] NEC Report *64 Bit, 25 MIPS SPARC,* in Nikkei Electronics 1989.4.3.

[4] M.Abe et. al. *Ultra-High Speed HEMT LSI Technology,* in Fujitsu Sci. Tech. J. 24.4. Dec 1989.

[5] Matsushita News *Japanese - English Interpreter,* Matsushita Electric Exhibition of Technology, Technical Report April, 1989.

[6] Y. Harafuji et.al. *C&C - NET LOOP 6560 400Mbps Optical Loop LAN for Multi-media Service in Intelligent Buildings* in NEC Tech J. Vol. 42 No. 2/1989.

[7] M. Hirose et.al. *Development of 410 Mbps Multi-media Highway* in National Convention Record of the Electronics Information and Communication Society April, 1989.

[8] Report of NTT International Symposium in Tokyo, 1988.

Lecture Notes in Engineering

Edited by C.A. Brebbia and S.A. Orszag

For information about Vols. 1-39 please contact your bookseller or Springer-Verlag.

Lecture Notes in Engineering

Edited by C.A. Brebbia and S.A. Orszag

Vol. 59: K. P. Herrmann, Z. S. Olesiak (Eds.)
Thermal Effects in Fracture
of Multiphase Materials
Proceedings of the Euromech Colloquium 255
October 31 - November 2, 1989, Paderborn, FRG
VII, 247 pages. 1990

Vol. 60: U. Schumann (Ed.)
Air Traffic and the Environment –
Background, Tendencies and Potential
Global Atmospheric Effects
Proceedings of a DLR International Colloquium
Bonn, Germany, November 15/16, 1990
VI, 170 pages. 1990

Vol. 61: A. Der Kiureghian,
P. Thoft-Christensen (Eds.)
Reliability and Optimization
of Structural Systems '90
Proceedings of the 3rd IFIP WG 7.5 Conference
Berkeley, California, USA, March 26-28, 1990
VIII, 405 pages. 1991

Vol. 62: Z. Zhao
Shape Design Sensitivity Analysis and Optimization
Using the Boundary Element Method
VIII, 192 pages. 1991

Vol. 63: H. A. Eschenauer, C. Mattheck,
N. Olhoff (Eds.)
Engineering Optimization in Design Processes
Proceedings of the International Conference
Karlsruhe Nuclear Research Center, Germany
September 3-4, 1990
XIV, 355 pages. 1991

Vol. 64: A. Elzein
Plate Stability by Boundary
Element Method
VII, 205 pages. 1991

Vol. 65: J. Holnicki-Szulc
Virtual Distortion Method
VIII, 176 pages. 1991

Vol. 66: M. A. Jabri
An Artificial Intelligence Approach
to Integrated Circuit Floorplanning
XIII, 149 pages. 1991

Vol. 67: S. Takahashi
Elastic Contact Analysis
by Boundary Elements
VI, 217 pages. 1991

Vol. 68: T. G. B. DeFigueiredo
A New Boundary Element Formulation
in Engineering
IX, 198 pages. 1991

Vol. 69: T. K. S. Murthy, C. A. Brebbia Eds.)
Advances in Computer Technology
and Applications in Japan
IV, 158 pages. 1991

Vol. 70: E. K. Bruch
The Boundary Element Method
for Groundwater Flow
II, 120 pages. 1991

GPSR Compliance

The European Union's (EU) General Product Safety Regulation (GPSR) is a set of rules that requires consumer products to be safe and our obligations to ensure this.

If you have any concerns about our products, you can contact us on ProductSafety@springernature.com

In case Publisher is established outside the EU, the EU authorized representative is:

Springer Nature Customer Service Center GmbH
Europaplatz 3
69115 Heidelberg, Germany

Batch number: 10704771

Printed by Printforce, the Netherlands